OUTDOOR GEOGRAPHY

ALSO AVAILABLE FROM LIVING BOOK PRESS

The Burgess Animal Book for Children (in color)
The Burgess Bird Book for Children (in color)
The Burgess Flower Book for Children (in color)
Home Geography—C.C. Long
Elementary Geography—Charlotte Mason
Viking Tales—Jeannie Hall
Parables From Nature*—Margaret Gatty
Fifty Famous Stories Retold*—James Baldwin
The Blue Fairy Book* —Andrew Lang
The Red Fairy Book —Andrew Lang
Beautiful Stories of Shakespeare* —Edith Nesbit
Tales from Shakespeare* —Charles and Mary Lamb
    * in AO reading order

Richard Halliburton's Marvels of the Orient
Richard Halliburton's Marvels of the Occident

Charlotte Mason's Home Education Series
1. Home Education
2. Parents and Children
3. School Education
4. Ourselves
5. Formation of Character
6. A Philosophy of Education
And many, many more!

All Living Book Press titles are complete and unabridged, and (when possible) presented with the original illustrations, sometimes from several sources, to bring these great books even more to life.

To see a complete list of all our releases or if you wish to leave us any feedback please visit www.livingbookpress.com

OUTDOOR
GEOGRAPHY

HERBERT HATCH

LIVING BOOK
PRESS

This edition published 2019
By Living Book Press
147 Durren Rd, Jilliby, 2259

Copyright © Living Book Press, 2019

ISBN: 978-1-925729-46-7

A catalogue record for this
book is available from the
National Library of Australia

Contents

PREFACE   I

INTRODUCTION   2

CORRELATION WITH ENGLISH LITERATURE   4

SECTION I
STUDY OF THE SKY

1. The Sun at Noon—I   6
2. The Sun at Noon—II   7
3. The Course of the Sun—I   8
4. Sunrise and Sunset   10
5. The Length of Daylight   11
6. The Course of the Sun—II   12
7. The Course of the Sun—III   14
8. The Course of the Sun—IV   15
9. Finding North by a Watch   16
10. The Sundial—I   18
11. The Sundial—II   19
12. Height of the Sun   20
13. Altitude of the Sun and Latitude   21
14. To Find the Size of the Earth by Observation of the Sun   22
15. To Find Latitude from the Sun   23
16. Representation of Sun, Moon,and Earth   24
17. Observation of the Moon—I   25
18. Observation of the Moon—II   26
19. Size of the Moon   27
20. Eclipses of the Moon   28
21. Observation of the Stars—I   29
22. Observation of the Stars II   30
23. To Find the Altitude of the Pole Star   31
24. Latitude from the Pole Star   32
25. Longitude—I   33
26. Longitude II   34
27. The Planets   34

SECTION II
WEATHER OBSERVATION

28. Heat and Cold   36
29. The Thermometer I   37
30. The Thermometer II   38
31. The Maximum and Minimum Thermometers   39
32. Rain—I   41
33. Rain—II   42
34. The Rain-gauge   43
35. Rainfall   45
36. Wind   46

37. Clouds   47
38. Wind Directions   48
39. Dew, Fog, and Mist   49
40. The Barometer   51
41. Weather Record I   53
42. Weather Record —II   53
43. Weather Record III   54
44. Weather Forecasting   57

SECTION III
PLANS AND MAPS

45. The Compass—I   58
46. The Compass—II   59
47. The Compass—III   60
48. Determination of Geographical North   62
49. Introduction to Plans   63
50. Introduction to Maps—I   63
51. Introduction to Maps—II   64
52. The Chain—I   65
53. The Chain—II   66
54. The Chain—III   67
55. Chain Survey Plan   68
56. Chain Survey (Area)   69
57. To Draw a Right-angle   69
58. Measurement of Width of River—I   70
59. Measurement of Width of River—II   71
60. The Globe and Maps   72
61. Lines of Longitude and Latitude   73
62. Size and Shape of the Earth   74
63. The Plane Table—I   75
64. The Plane Table— II   76
65. The Plane Table—III   77
66. Setting an Ordnance Map   78

SECTION IV
HEIGHTS AND CONTOURS

67. Heights   80
68. Bench-marks   81
69. Determination of Heights—I   83
70. The Clinometer   84
71. Determination of Heights— II   86
72. The Water-level   87
73. Gradients   88
74. Determination of Heights—III   89
75. Determination of Heights— IV   90
76. Contour Lines—I   91
77. Contour Lines—II   91
78. Contour Lines—III   92

SECTION V
LAND AND SEA

79. Water 94
80. Springs 96
81. Action of Frost 97
82. Rivers—I 98
83. Rivers—II 99
84. Waterfalls 100
85. A Rainbow 102
86. Rocks—I 103
87. Rocks—II 104
88. Minerals 105
89. Soils 106
90. Sea-water and Waves 107
91. Tides 109

SECTION VI
HUMAN GEOGRAPHY

92. Roads 112
93. Railways 114
94. Canals 115
95. Local Town Industries 116
96. Local Farming 117
97. Pleasure and Health 118
98. Historical Geography 119
99. Local and State Government 120
100. Comparison with other Districts 121

APPENDIX I

Other Useful Apparatus 125

APPENDIX II

Maps 126

APPENDIX III

Table of Declination of Sun 127

APPENDIX IV

Solar and Clock Time 128

APPENDIX V

Magnetic Declination 129

PREFACE

The author has for many years taught Practical Geography to classes of teachers, and has found that there is a general feeling that outdoor work with the children is very desirable. The difficulty has been that such lessons have often been somewhat aimless and out of touch with indoor work.

The present book has been written to suggest a course of work which it is hoped will lead to observation and thought likely to make classroom teaching both easier and more fruitful. Each teacher must choose for himself the order in which he will give the lessons. It will depend upon the weather, the season, the situation of the school, and other circumstances. No attempt has been made to indicate even the age of the children who should attempt a certain exercise. An experienced teacher will judge of this best by his knowledge of the capacity of his pupils, and many of the exercises can be made easy or difficult as the teacher may wish. Some lessons should be repeated, e.g. Lesson 12 should be given several times at different seasons. There will probably, therefore, be enough material in the book for an outdoor lesson once a fortnight, on the average, from Standard III to the top of the elementary school. The lower forms of the secondary school should work the harder exercises, some of them perhaps in a more advanced way.

The author is indebted to Miss F. C. Cliff, B.A., for help in choosing the quotations from English literature, to Mr. A. J. Fawthrop, B.Sc., for many valuable suggestions and criticisms, and to Mr. A. I. Burnley for general assistance, particularly with the illustrations.

1

INTRODUCTION

Perhaps the best definition of Geography is "the study of the earth in its relations to man".

It is obvious that, in order to obtain a real knowledge of the earth, the child must observe for himself. Books and oral descriptions only give him second-hand knowledge. Maps, models, and sections provided for him are purely conventional and artificial, and cannot be properly understood unless he has constructed similar ones from nature.

Such observation is carried on to some extent even by the dullest child on his own account. Outdoor lessons, however, will be much more effective, because the teacher can call attention to the important points and direct the child's mind along the lines desired. The interest in geography so caused will lead to many valuable observations being made by the child when the teacher is no longer with him.

Books and maps will then have a much fuller value, because the knowledge they give is dependent upon the knowledge of the real world that the pupil possesses before he uses them. Outdoor work is therefore particularly valuable in the elementary school.

It is as impossible to learn geography well from books and maps only, as it is to learn chemistry or botany from books and diagrams only. In both cases the value of such artificial aids comes in after some knowledge is obtained from experience of the *real things* which we are studying.

Many exercises in "practical geography" books consist of work on section-drawing from maps, drawing curves from statistics, &c. Such work may be valuable, but it should come *after* more truly practical observations based on facts themselves and not on artificial representations of facts.

Outside the classroom lies the great world of men. It is that which we have to study, and the best way to commence is to observe the part of it that we can reach. A knowledge of the regions beyond must be founded by comparison and contrast on what we can learn by the evidence of our senses.

CORRELATION WITH ENGLISH LITERATURE

There may be a danger of the treatment of geography becoming too cold and mathematical. If such a danger exists it arises from the teacher and not from the subject. Certain branches of geography should be studied in the same way as science or geometry, but the poetic and romantic view of the subject also exists and should be given due prominence.

The following extracts have been chosen because of their bearing on outdoor geography. In the English lesson they should be discussed from other points of view. For instance, referring to the extract after Chapter 4, it should be explained that the conversation is between conspirators plotting an assassination. While their leaders are speaking aside, the others, no doubt in a state of nervous suspense, are talking, not of the coming tragedy which fills their minds, but of trivial matters in order to pass the time.

It is suggested that the extracts should be learnt by heart by the children after the lesson to which they relate has been given.

SECTION I
STUDY OF THE SKY

Astronomy is difficult to teach, even in the most elementary way, to children, because it is necessary to consider the earth from a distance as it were. A conception of the solar system, with its central sun and the earth and other planets revolving round it, is difficult to understand, because we always see the heavenly bodies from a point within the system.

It is best, therefore, until somewhat advanced work is possible, to consider the sun, moon, &c., *as we see them*. It is unnecessary to interfere with the child's original idea that the earth is flat. This guide is considered to be a great hemispherical dome, with the sun, moon, and stars wheeling around it in great curves.

According to the age and capacity of the children, some idea of the almost infinite distances of some of the heavenly bodies should be given. If an express train (60 miles an hour) could leave the earth for the sun and travel night and day, then if it started when a child was born it would not have covered half the distance when he was an old man, 80 years afterwards. Yet the distance of the *nearest* fixed star is enormously greater than this. If the sun's distance from the earth be represented by 1 ft., the nearest star on the same scale would be over 50 miles away. Other stars are so much farther distant that all known methods of measurement have failed.

No lesson on the sky is complete which does not call attention to its beauty and grandeur. In the ugliest manufacturing or urban district teachers will here have an opportunity of developing the artistic sense of the children.

I. The Sun at Noon—I

The first facts to be taught about the sun are that it is a globe of fire, of enormous size and at an enormous distance. Then the fact that it is in the south at noon is the starting-point for a consideration of direction.

Smoke some pieces of glass by holding them over a lighted candle or taper. Let the children look at the sun through these glasses. The idea that it is quite small may be removed by observing it when setting behind distant trees or buildings, when it will quite dwarf them.

Make a clear mark on the school playground. Let a child stand here at noon (remember that 1 p.m. summer time is really noon), and mark the position of the shadow of his head. Tell the children that "south" means towards the sun at noon. In other words "north" means the direction where shadows point at noon. Let the children walk south, i.e. towards the sun, and north, i.e. away from the sun.

Ask what buildings, hills, &c., lie to the north and to the south of the school. Let the children face north with their arms stretched out to the side. Tell them that their right arms point east and their left arms west. Show that the reverse is true if they face south. Let the class march east and west.

On another day let the same child stand on the same mark at noon. Make it clear that the shadow always points the same way at this time.

Some direction exercises may be done in the class room, but it is more important to know direction outdoors than indoors. The distance and size of the sun can only be indicated in vague terms—figures are of no use.

I am the Angel of the Sun,
Whose flaming wheels began to run
When God's almighty breath
Said to the darkness and the night:
"Let there be light", and there was light.

Longfellow.

2. The Sun at Noon— II

The last lesson may now be made more definite and a meridian may be drawn.

By indoor experiments, with a candle or lamp and a rod fixed in an upright position on a board, show that the shadow always points away from the light and that the length of the shadow depends upon the height of the lamp. A clear understanding of these points is essential.

Choose a place in the yard which will be in sunshine most of the day. Place a straight stick (5 or 6 ft. long) in an upright position. Mark the end of its shadow at various times, some in the morning, some in the afternoon. Use Greenwich time, remembering that summer time is one hour later, e.g. 12 noon Greenwich is 1 p.m. summer time.

Take a long string, tie it to the bottom of the pole, and at noon (Greenwich) pull it out so that it lies along the shadow but stretches farther. In this way the noon shadow can be drawn longer. Draw the line. Young children may call this the "noon line". Older children will be taught the name "meridian", or "line of longitude", and the fact will be pointed out that some noon lines are marked on most maps. If lengthened, all noon lines would reach the north and south poles. Older children will understand that all meridians, within a mile or so, are practically parallel, meeting at a very great distance, and that there are an infinite number of them. If possible, a "noon line" should be drawn in the class room.

Let a child try to tread on his shadow at noon. As the shadow is to the north of him, he will be obliged to walk to the north.

NOTE 1.—Solar noon, even at Greenwich, may be as much as fifteen minutes different from clock noon. Also every degree of longitude makes a difference of four minutes in solar time. See Appendix IV. Of course, solar noon may be found by noting the time when shadows are shortest.

NOTE 2.—*Any* line drawn due north and south is part of a meridian, and not merely those which are an exact number of degrees east or west of Greenwich.

> I feed the clouds, the rainbows, and the flowers
> With their aethereal colours ; the Moon's globe
> And the pure stars in their eternal bowers
> Are cinctured with my power as with a robe:
> Whatever lamps on Earth or Heaven may shine,
> Are portions of one power, which is mine.
>
> Shelley.

3. The Course of the Sun—I

Some idea of the movement of the sun in the sky may now be given.

The teacher must judge for himself whether the observations can be made best by the children on a Saturday, or whether a few minutes several times a day can be spared for the purpose during school-time.

In either case the child must have a fixed position from which all the observations must be taken. A chalk circle large enough for the child's feet may be marked out. For direction it will be sufficient to say "over the market-place" or "just to the right of — Hill", &c.

Questions such as the following should be considered, and next day a lesson should be devoted to talking them over and answering them:—

1. Where was the sun at 9 a.m. ? [A place about S.E.]

2. Where was the sun at noon?[A place about S.]

3. Where was the sun at 3 p.m. ? [A place about S.W.]

4. When was the sun over X (a place due north)? [Never.]

5. Did the sun appear to move to your right or to your left as you faced it ? [Right.]

6. When was it highest in the sky ? [Noon.]

7. Was it ever directly overhead ? [No.]

8. At noon was it nearer the ground or the overhead point (zenith)? [If summer, point; if winter, nearer ground.]

It may now be pointed out that we sometimes think that an object is moving when it is still and we are moving the opposite way. Refer to apparent movements when in a train. Tell the children that this apparent movement of the sun is due to the earth moving round like a top. Point out that as we look southwards the sun seems to move from left to right, therefore we must be moving from right to left, i.e. from west to east.

> I stand at noon upon the peak of Heaven,
> Then with unwilling steps I, wander down
> Into the clouds of the Atlantic even.
>
> Shelley.

4. Sunrise and Sunset

The next step is to observe when and where the sun rises and sets. For young children it is enough to notice that sunrise is towards the east and sunset towards the west. The points are exactly east and west, however, only on 21st March and 21st September. Older children should notice that in summer both rising and setting occur nearer the north, and in winter nearer the south.

In a period of fine, settled weather the children should be encouraged to notice the exact points of rising and set-

Date.	Place of Rising.	Place of Setting.	Time of Rising.	Time of Setting.
About Mar. 21	About E.	About W.	About 6 a.m.	About 6 p.m.
,, June 21	,, N.E.	,, N.W.	,, 4 a.m.	,, 8 p.m.
,, Sept. 21	,, E.	,, W.	,, 6 a.m.	,, 6 p.m.
,, Dec. 21	,, S.E.	,, S.W.	,, 8 a.m.	,, 4 p.m.

ting from a known place of observation. Such observations must be made at each season of the year. The results may be tabulated as follows:—

The exact places and times will vary according to the position of the school.

In a formal lesson the class may be taken to the place of observation and the approximate path of the sun pointed out. It will then be clear that the sun has farther to go in summer, so that it must then be longer above the horizon.

Decius. Here lies the east: doth not the day break here?
Casca. No.

Cinna. O, pardon, sir, it doth ; and yon grey lines
> That fret the clouds are messengers of day.

Casca. You shall confess that you are both deceiv'd.
> Here, as I point my sword, the sun arises;
> Which is a great way growing on the south,
> Weighing the youthful season of the year.
> Some two months hence up higher towards the north
> He first presents his fire.
>
> Shakespeare.

5. The Length of Daylight

Previous to this lesson, observations should be taken on the time of rising and setting of the sun. Should the sun set behind a hill, allowance must be made for this. Point out that darkness comes more quickly than it would if the hill were not there. Tables showing sun rise and sunset are common enough, and if the school is far north of London it may be shown that the sunset times are later in summer and earlier in winter than at London.

If possible, at least one observation a month should be made during a whole year.

By means of the table showing the hours when the sun is above the horizon at London, draw a graph similar to the one shown on the following page. The side of one small square may represent 30 minutes, or a still larger scale may be used.

Point out that in more northerly latitudes the line will be more curved, i.e. it will be lower in January and December and higher in midsummer. In any place, however, the total hours of day in a year will equal the total hours of night. This is shown by the fact that the shaded portion below the dotted line (at twelve hours) is equal to the unshaded portion above that line.

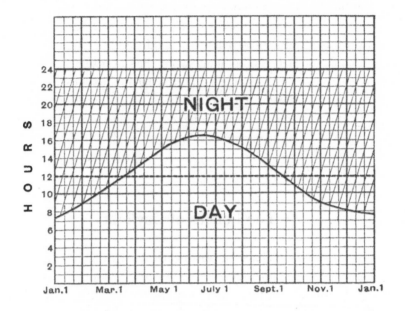

DURATION OF DAYLIGHT AT LONDON

		Hrs.	Mins.			Hrs.	Mins.
Jan.	1	7	50	July	1	16	39
Feb.	1	9	5	Aug.	1	15	31
Mar.	1	10	50	Sept.	1	13	40
April	1	12	54	Oct.	1	11	41
May	1	14	51	Nov.	1	9	4
June	1	16	23	Dec.	1	8	8

And still with laughter, song, and shout
Spins the great wheel of earth about.

Stevenson.

6. The Course of the Sun—II

The experiment described in Lesson 2 should now be worked more carefully, using a stick 6 to 8 ft. long and marking the end of its shadow every half-hour, if possible, from 9 a.m. to 4 p.m. Greenwich.

At the same time the experiment may be done on a small scale, using a long needle fixed vertically in a sheet of paper on a drawing-board. Care must be taken to mark the exact position of the board, so that if it is accidentally moved it can be replaced. —

The following questions should be considered :

1. Which way does the 9 a.m. shadow point? [N. W.]

2. In what part of the sky is the sun at 9 a.m.? [S. E.]

3. Judging from the shadows, where would the 6 a.m. shadow point? [West.]

Tell the children that the sun is always due east at 6 a.m. and due west at 6 p.m. In winter it is below the horizon at these times, and in summer it is well above the horizon. It is evidently inaccurate, therefore, to say that the sun rises in the east and sets in the west.

4. Where does it seem that the sun would be at midnight? [North.]

Explain that in this country the sun is always below the horizon at midnight and therefore we never see it in the north.

5. At what time do the shadows stop growing shorter, and begin to lengthen? [At noon.]

6. At what time of day is the sun highest? [At noon.]

7. At what time of night is it lowest? [At midnight.]

8. When in the whole year is it highest? [Noon at midsummer.]

9. When in the whole year is it lowest? [Midnight at midwinter.]

Explain how the sun lights up the sky when it is not far below the horizon, so that there is twilight for about an hour before sunrise and after sunset. Also the summer nights in England, and still more in Scotland, are not pitch dark at all unless cloudy.

7. The Course of the Sun—III

The observations of Lesson 6 should be made more definite by a model.

Take a large circular piece of cardboard to represent the district around the observer, whose position is supposed to be at the centre. Draw on it a rather smaller concentric circle. Prepare another disc of cardboard of the same size as this drawn circle. On the diameter of the latter cut a slit and push the smaller cardboard circle halfway through this slit.

If the small circle is inclined to the larger at the correct angle its circumference will represent the apparent part of the sun on 21st March and 21st September. The angle should be 90°, less the latitude of the place, i.e. it will equal the angle of the sun's altitude at noon on the dates mentioned.

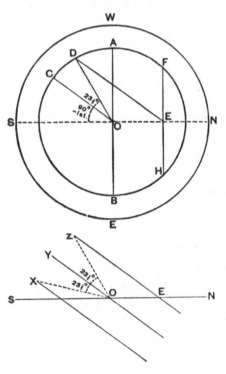

Now cut two other slits in the large disk, parallel to the first. They must be obtained by the construction shown in the diagram. Prepare two more cardboard circles of the same size as the lesser one previously made. Push one through a slit so that the greater portion of it is above the slit. Its circumference, if it is inclined at the same angle as the other one, will represent the sun's path on 21st June. Similarly, the third circle may be fixed to show the path on 21st December. The upper

diagram is the plan of the large cardboard disk. AB is the
first slit, at right angles to SN. OC is drawn at an angle to SN
equal to 90° diminished by the latitude of the place. COD is
23 ½°. DE is drawn parallel to CO. Through E is drawn FH
parallel to AB. FH is the second slit. The third is drawn the
same distance to the left of AB that FH is to the right.

The lower diagram is the end elevation of the model. X, Y,
and Z are the highest points of the disks, i.e. they represent
the noon positions of the sun at midwinter, the equinoxes
and midsummer respectively.

8. The Course of the Sun—IV

The model should be taken outdoors on a sunny day. A
drawing-pin may be fixed on the edge of one of the disks to
represent the sun.

If circumstances have prevented a model being made, a
diagram such as that shown may be substituted for it. The
ellipse SWNE shows the district round O, the point of observation, seen in perspective. As before, XYZ are the noon
positions of the sun at the various seasons.

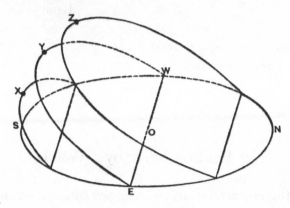

Whether the model or the diagram is used, the facts in the previous lessons should be made more clear.

The children must quite grasp the fact that the sun appears to move in a circle, which is tilted always at the same angle for any particular place of observation. It should be next explained that this angle becomes greater as we travel south, till at the equator the sun is vertically overhead at noon on 21st March and 21st September.

Similarly, in places farther north the sun's path is tilted less, so that the sun rises and sets nearer the north in summer and nearer the south in winter. This gives longer summer days and shorter winter ones. If we go as far north as "the land of the midnight sun", we find that the sun fails to set at all on some midsummer days and fails to rise at all on some midwinter nights.

Show that the midnight sun is always in the north in our hemisphere. Show how long the twilight is in northern latitudes and how short in the tropics.

See the following quotations. The first one refers to our own latitudes, the second to tropical regions.

Twilight and evening bell,
And after that the dark.

Tennyson.

The Sun's rim dips; the stars rush out;
At one stride comes the dark.

Coleridge.

9. Finding North by a Watch

Let the children draw a circle, say of 6 in. diameter, on a sheet of cardboard and divide it into twenty-four equal parts. Make a movable pointer, fastened by a drawing-pin at the centre of the circle. At the end A put a vertical needle.

This may be considered as a watch face, but divided into twenty-four hours instead of twelve as an ordinary watch. The pointer stands for the hour hand. Set the pointer to the correct time of day.

Take it into a sunny position and direct the pointer towards the sun. This will be easy, for in the proper position the shadow of the needle at A will fall along the pointer AB. The point marked noon will now be south, that marked 6 a.m. will be east, and so on.

The children should have a clear idea of the sun going round the sky in a huge circle, part, of course, below the horizon, and tilted at an angle to the horizon. If so they will have little difficulty in understanding the above.

It may then be used as a simple sundial. It must be fixed so that the line joining the noon and midnight marks runs from south to north, and then when the pointer is turned to the sun it will show the time.

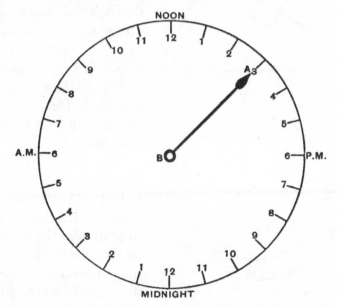

Explain that the circle on the cardboard represents the huge circle which the sun describes in the sky in twenty-four hours. The hour hand of a watch, however, describes a circle in twelve hours, i.e. half the time that the sun takes to complete its circle. As the watch hand moves twice as quickly, we can use it as we do the above cardboard dial but we must halve the angle.

The rule, therefore, is as follows. Point the hour hand of the watch to the sun. Bisect the angle between the hour hand and the XII mark (or the I mark in summer time) and the bisecting line points south.

10. The Sundial—I

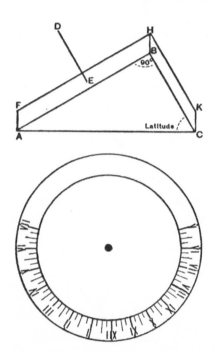

In the woodwork lesson let the boys prepare a pieceof wood as shown. The face ABHF should be about 3 in. square. The other faces do not matter, but the angle BCA must equal the latitude of the school. The boys should find it from an atlas or wall-map. CBA is a right angle.

On a piece of paper let them draw two circles with radii 2½ in. and 3 in. Divide each circle into twenty-fourths, for the hours of the day, and these divisions into quarters. Of course the night hours need not be marked. Gum the paper on to the face ABHF so that the XII mark is lowest. In the centre of the paper fasten a stout needle, taking care that it is exactly perpendicular to the sloping face. Fix the sundial so that the face ABC (and consequently a line from the needle to the XII mark) points exactly north.

11. The Sundial—II

At half-hour intervals during the whole of a sunny day let the dial be compared with a clock. Record the results in the manner shown below.

Let children bring information respecting sundials they know of on neighbouring churches, &c. They should be told to be able to answer the following questions:—

1. Is the dial vertical, horizontal, or neither?
2. If vertical, which way does the wall face? [South.]
3. Why is this? [To have the sun almost all day.]
4. In what direction does the pointer (gnomon) lie? [N and S.]
5. What hours are omitted from the dial? [Early and late.]
6. Why is this? [The sun will not be shining.]
7. What angle does the gnomon make to the horizontal? [Equal to latitude.]
8. To what point in the sky does the gnomon point? [To the Pole Star.]
9. What motto is on the dial?

The lesson should be devoted to considering the above and discussion as to the weakness of the dial. Tell the class that the earth does not move always with the same speed, and therefore the sun does not apparently move at the same speed. Dial time, therefore, is sometimes fast and sometimes slow, like sun time is. Our clocks are regulated to mean or average sun time, as of course it would be difficult to make them hurry some times and slow down at other times to keep with the sun.

Clock Time.	Dial Time.	Dial Fast or Slow by Clock Time.
9 a.m.	9.5 a.m.	Five min. fast.
9.30 a.m.	Sun obscure.	———
10 a.m.	10 a.m.	Correct.
&c.	&c.	&c.

12. Height of the Sun

To find the angle which the sun's rays make with the ground.

Let one of the children measure the length of the shadow of a vertical pole in the playground at noon. Let others measure the shadow of a vertical needle (see Lesson 6). The length of the A pole and of the needle must also be accurately measured.

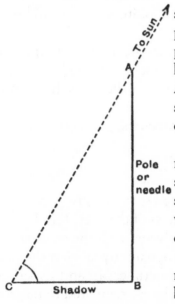

Make a drawing to scale as shown, preferably on squared paper. If AB is the length of the pole or the needle and BC is the length of shadow, then the angle ACB will give the angle to the sun, or what is called the altitude of the sun.

This experiment should be repeated, say every month, to show how the altitude of the sun at noon increases from midwinter to midsummer, and then decreases.

The approximately correct results for latitude 51 ½° are given below. One degree must be subtracted for every degree that the school is north of 51½°, and one degree added for every degree south of that latitude.

Date.	Altitude.	Date.	Altitude.
January 1.	$15\frac{1}{2}°$	July 1.	$61\frac{3}{4}°$
February 1.	$21\frac{1}{4}°$	August 1.	$56\frac{1}{4}°$
March 1.	$30\frac{1}{2}°$	September 1.	$47°$
April 1.	$42\frac{3}{4}°$	October 1.	$35\frac{1}{2}°$
May 1.	$53\frac{1}{4}°$	November 1.	$24\frac{1}{4}°$
June 1.	$60\frac{1}{2}°$	December 1	$16\frac{3}{4}°$

13. Altitude of the Sun and Latitude

It is first necessary to show that the altitude of the sun varies inversely as the latitude, i.e. if the latitude of A is 3° more than that of B, then the sun's altitude at A at a certain time will be 3° less than the sun's altitude at B.

Draw a large circle in the playground to represent the earth. If drawn with a radius of 33 ft. it will be on the scale of ¹/₁₀ in. to a mile (see Lesson 62).

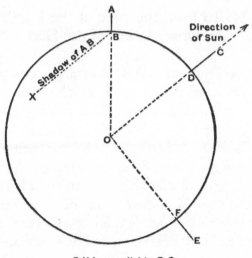

B X is parallel to D O

Explain that by "down" we mean towards the centre of the earth. Objects fall towards the centre of the earth, and a vertical line is one drawn towards that point. AB, CD, EF represent vertical lines, and they are not parallel.

If the sun is in the direction shown on the diagram, a pencil erected at D will throw a shadow pointing to wards O. This represents then a place where the sun is vertical. Show that a pencil erected at F (90° from D)throws a shadow on the line which represents the ground. F then represents a place where the sun's altitude is 0°.

Moving 90° over the earth's surface has therefore changed the sun's altitude by 90°.

At B the number of degrees from a vertical sun is given by the angle BOD. The shadow there is not vertical but at an angle XBO to the vertical. But the angle XBO and BOD are equal.

The above is difficult for children, but they will under stand that as we move away from a place where the sun is vertical the direction to the sun becomes less and less vertical.

14. To Find the Size of the Earth by Observation of the Sun

For this determination it is necessary to compare results with some other school far to the north or south. Suppose two schools, at Birmingham and Plymouth respectively, agree to work together. On a certain day the altitude of the sun at noon must be found from each place. It will probably be better to agree on several days, lest some should be cloudy.

Suppose the results are: Birmingham 55°, Plymouth 57¼°. Find from the map the distance that one town is north of the other. It is about 160 miles. Work out a proportion sum. As the difference in altitude is 2¼°, and this corresponds to 160 miles, a 160 difference of 1° would correspond with 160÷2¼= 71 miles.

The circumference of the globe is 360°. Therefore in miles it must be 360 × 71 = 25,560 miles. Correct equatorial circumference = 24,872 miles.

15. To Find Latitude from the Sun

Find the noon altitude of the sun as in Lesson 12. Subtract this from 90° to find how many degrees the sun is from a vertical position.

Starting from the place where the sun is directly overhead, every degree moved in distance will mean that the sun will be a degree lower in the sky. Therefore the number of degrees of your school from where the sun is overhead can be found.

For example, suppose that the altitude of the sun at noon is 20°. Then we are 90° − 20° or 70° from an overhead sun.

We must now find where the sun is overhead. This will be at 23½° N. (Tropic of Cancer) on 21st June, 23½°S. (Capricorn) on 21st December, and 0° (Equator) on 21st March and 21st September. The latitude where the sun is directly overhead on any particular day is called its declination. (See Appendix III.)

Suppose the experiment is worked on 15th February. Then if we are 70° from an overhead sun, we are 70° from 13° S. Therefore we are 70°− 13° or 57° from the equator, i.e. in latitude 57° N.

Take another example. Suppose that on 15th August the altitude was 52¾°. Then we were 37¼° away from an overhead sun, i.e. 37¼° away from 14¼° N., i.e. 51½° N. (or of course 23° S.).

16. Representation of Sun, Moon, and Earth

Book diagrams always give wrong ideas of the relative distances and sizes of the heavenly bodies. This is because it is impossible to draw to true scale on an ordinary sheet of paper. The following will give a true idea.

Take a globe of 1 ft. diameter for the earth—this is the size of the ordinary school globe.

On the same scale the moon will be a globe of 3½ in. diameter. A tennis ball will do for this and it should be 30 ft. away from the model earth.

Tell the class that the sun should be represented by a globe 100 ft. in diameter and 2¼ miles away. Mark on the wall the approximate size of such a globe, and mention a place roughly the right distance for it to be placed.

The movements of the moon and earth can be represented. The globe (earth) will be moving in a circle round the large globe (sun) which is 2¼ miles away. Therefore the path of the model earth will be about 14 miles long (really the earth's orbit is elliptical, but it is so nearly circular that the difference is negligible).

The moon will be moving round the earth. It does not move quite in the same plane as the earth. This can be shown by keeping the earth globe always the same distance from the ground, while the tennis ball is sometimes higher and sometimes lower.

Thou art speeding round the sun,
Brightest world of many a one;
Green and azure sphere which shinest
With a light which is divinest
Among all the lamps of Heaven
To whom life and light is given.

Shelley.

17. Observation of the Moon— I

Let the children have a piece of paper containing thirty or thirty-one circles the size of a halfpenny. During the whole of one month let them fill in the circle to show the moon's shape, e.g. fill in all the circle when the moon is full, and so on. Below each circle let them put the time of day or night when they saw the moon.

In the lesson, consider the following:—
1. How long is it from full moon to full moon again? [Four weeks].
2. Is the moon seen every clear night? [No.]
3. When the moon is seen as a thin crescent, at what time in the evening is it seen? [Soon after sunset. Also a little before sunrise for the waning moon, but this will probably not be seen.]
4. Is the moon ever seen in the daytime? [Yes.]
5. Why does it then look so pale? [Because of the strong sunlight. Compare with candle in daytime and at night.]

Some of the most important points about the moon may be explained, e.g. that it is somewhat like our own earth but smaller; that it is very distant but nearer than the sun or stars; that it shines because the sun shines on it and lights it up. Compare with clouds which look dark in the shadow but bright when the sun shines on them. " Every cloud has a silver lining."

In broad daylight, and at noon,
Yesterday I saw the moon
Sailing high, but faint and white
As a schoolboy's paper kite.

Longfellow.

18. Observation of the Moon—II

This lesson should be preceded by observation for a whole month. During that time a diary should be kept like the one below.

Date.	Drawing of Moon.	Time of Rising.	Time of Setting.	Time when Moon was seen.	Direction where seen.
Nov. 13	☽		{ Near setting at 10 p.m.	} 10 p.m.	West.
,, 14	} Cloudy				
,, 15					
,, 16	☾				
,, 21	○		7.30 a.m.	7.30 a.m.	{ A little N. of W.
,, 24	☾	10.30 p.m.		8.45 a.m.	

With the aid of this diary answer the following questions:—

1. If the moon has a certain shape on one day, how many days will elapse before it has the same shape again? [Twenty-eight.]

2. If you are facing the sun, whereabouts will the full moon be, if both are in the sky? [Behind you.]

3. When the new moon is seen, is it close to the position of the sun or far away from it? [Near the sun.]

4. When the moon is waxing (i.e. growing bigger), do the points of crescent point to the left or right? [Left.]

5. Answer the same question for a waning moon. [Right.]

6. Where and when can the waxing new moon be seen? [West about sunset]

7. Where and when can the waning crescent moon be seen? [East about sunrise.]

Often when the moon is young the rest of the disk may be faintly perceived. Explain that this dim portion has "earth-light" falling on it, for if the living creatures are on the moon the earth will appear to them as a very large moon. This faint "earth-light" makes the surface visible to us.

Then the moon in all her pride,
Like a spirit glorified,
Filled and overflowed the night
With revelations of her light.

Longfellow.

19. Size of the Moon

Some conception of the size and distance of the heavenly bodies should be given, and some idea of how they are found.

The following experiment can be done at home and gives remarkably accurate results. Cut from of paper piece a circle 1 in. in diameter and paste it on a window in sight of the full moon. Find how far away the eye must be from this paper disk in order that it may just hide the moon. A second person should measure the distance with a piece of string.

The calculation may be worked out at school by the following proportion sum: —

Distance of eye from disk : 240,000 miles : : width of disk : diameter of moon.

As three of these four numbers are known, the last can be found. The actual diameter is 2160 miles.

A similar experiment may be tried out of doors with a much larger disk held by one boy and observed by another boy several feet away.

Tell the children that the distance of the moon is obtained by taking a long base line (hundreds or thousands of miles long) on the earth and obtaining the angle to a point on the moon from the two ends at the same moment.

Art thou pale for weariness
Of climbing heaven and gazing on the earth,
Wandering companionless
Among the stars that have a different birth,—
And ever changing, like a joyless eye
That finds no object worth its constancy?

Shelley.

20. Eclipses of the Moon

Take outside on a sunny day a globe 1 ft. in diameter to represent the earth, and a ball about 3 ½ in. diameter to represent the moon. Of course any other sizes will do, provided that the large ball has a diameter about three times that of the small one or rather more. It will be better if you can hang the little ball by a string. Take it round the big one at a distance of about 30 ft., and it will represent the moon's revolution round the earth in one month. Show that the illuminated part of the moon will sometimes be turned towards the earth and sometimes away from it. In the former case we have full moon, and in the latter new moon, i.e. the moon is invisible to us.

Let the class consider the following questions: —

1. May the moon go round without entering the earth's shadow, i.e. without being eclipsed? [Yes.]

2. If it is eclipsed, will it be at new moon, full moon, or what phase? [Full.]

3. Over what part of the earth will the eclipse of the moon be visible? [Everywhere that the moon is visible.]

4. May the moon's shadow be thrown on the earth, thus causing an eclipse of the sun? [Yes.]

5. If so, will it happen at full moon, or what other phase? [New moon.]

6. Over what part of the earth will the eclipse of the sun be visible? [A small part.]

The surface or plane joining the sun and the path of the earth is called the ecliptic. Notice that it is only when the moon is in the same plane that eclipses occur, and this is why the ecliptic is so named. If the moon is not in the ecliptic the shadows are thrown into space, and neither moon nor earth is affected.

> Till clomb above the eastern bar
> The horned Moon, with one bright star
> Within the nether tip.
>
> Coleridge.

21. Observation of the Stars— I

The children should be told to look at the sky on a fine winter evening when the moon is not visible.

The following questions should afterwards be asked:—
1. How many stars are there? [They cannot be counted.]
Point out that some stars are only just visible. A telescope will show many that our eyes alone cannot see, and the stronger the telescope the more stars can be seen. So far as we know, there is no end to the number of the stars.
2. Are they all of the same brightness? [No.]
3. What may cause this? [Some being bigger or nearer.]
Tell the class that the stars are like our sun but much farther away. Some are bigger than our sun and some less, but because of their distance they look much smaller. Compare with gas lamps on a distant hill.

4. Is there any heat from the stars? [No.]

This is because of the great distance. A lamp some miles away can be seen but no heat can be felt from it.

5. Are there any stars that do not twinkle? [Yes.]

With such young children it is best simply to tell them that these stars are not so far away. Without using the word planet they may be told that these stars are like our earth and perhaps have living creatures on them.

> Continuous as the stars that shine
> And twinkle on the milky way.
>
> Wordsworth.

22. Observation of the Stars II

On another fine evening let observations be made to answer the following questions. The place of observation must be the same throughout, and it is better to look northward.

1. Did any stars move during the evening? [Yes.]

2. Did any stand still all the time? [A few. Really the Pole Star only, but others near it move very little in a few hours.]

3. Did the distance between any two alter? (Best measured by holding up a pencil at arm's length as in model drawing.) [No.]

4. What might cause this apparent movement? [Either movement of sky or of earth.] (Refer to apparent movement of trees, houses, &c., when observer is in a train.)

Let the children copy a drawing of the Plough on a piece of paper, and notice its appearance as they turn the paper round. Let them take the papers home, so that they can find the Plough and the Pole Star for themselves.

I find my zenith doth depend upon
A most auspicious star.

Shakespeare.

Look how the floor of heaven
Is thick inlaid with patines of bright gold.

Shakespeare.

23. To Find the Altitude of the Pole Star

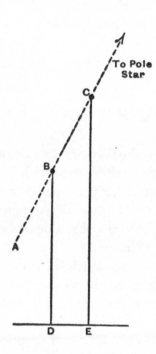

This must, of course, be done out of school hours, but a preparatory lesson should be given.

Draw the Great Bear (Charles's Wain or the Plough) and explain why these names are given to it. Explain the following way of finding the altitude:—

1. Place a small stick in a horizontal position (shown in end elevation at C), and as high above the ground as possible. Have a long string tied to it, so that its height can easily be found.

2. Have another stick, B, exactly similar but much lower in position.

3. It will be necessary to arrange these parallel sticks so that they are both in a line with the Pole Star when viewed from A.

4. The distance between the ends of the strings on must then be measured along the ground. At school the diagram can be drawn to scale, preferably on squared paper. The scale

should be large, say 1/10 in. DE will be drawn first, then BD and CE, and the inclination of CB to DE found. This will equal the altitude of the star. If correct, it will equal the latitude of the place (see next lesson). The altitude of the Pole Star may also be found by using the clinometer.

> But I am constant as the northern star,
> Of whose true-fix'd and resting quality
> There is no fellow in the firmament.
> The skies are painted with unnumber'd sparks;
> They are all fire, and every one doth shine;
> But there's but one in all doth hold his place.
>
> Shakespeare.

24. Latitude from the Pole Star

The class should have observed the Pole Star (see Lesson 22) and found its altitude (i.e. angle to the horizontal).

Take the globe into the playground. Let the class point out where the Pole Star would be seen at night. Turn the globe so that the axis points to this point in the sky. The globe is now parallel to the real position of the earth.

Make it clear that *downwards* means towards the centre of the earth, and *upwards* just the opposite direction.

1. Where would the Pole Star appear if you were at the North Pole? [Straight overhead.]

2. Where if you were at the Equator? [Close to the ground.]

Let the children look at their atlases or a wall-map and see that the equator is called latitude 0° and the North Pole latitude 90°. The latitude of a place is shown by the altitude of the Pole Star.

3. What is the altitude of the Pole Star from your town?

Find the town in the atlas and see what latitude is given there. It should agree with the altitude.

4. What should we notice about the Pole Star if we travelled northwards? [It would mount higher in the sky.]

25. Longitude— I

Insert a needle into a globe and take it into the sunlight. Notice the shadow of the needle on the globe. Rotate the globe. When the needle is opposite the sun the shadow will be the shortest possible. Point out that when the globe has made a complete revolution, i.e. after what represents twenty-four hours, the shadow is again the least possible. Explain how the same thing happens with our own earth. The shortest shadows are at noon.

Refer to the terms *meridian*, *noon line*, and *line of longitude* (see Lesson 2), and show the one marked on the playground. Those usually drawn on a map are numbered from 0 to 180 W. and 0 to 180 E., starting from London. 180 E. and 180 W. are the same line.

As it takes 24 hours for these 360 lines to pass in front of the sun, one must pass every $24/360$ hour $= 1/15$ hour $= 4$ minutes.

Therefore noon comes to London 4 minutes before it comes to a place 1° W., and 4 minutes after it comes to a place 1° E.

Numerous examples should be given, e.g.:

If it is 10 a.m. at London, what time is it at—

Liverpool, 3° W.? [9.48 a.m.]
New York 75° W.? [5 a.m.]
165° W.? [11 p.m. the day before.]
Calcutta 90° E.? [4 p.m.] &c.

What time should it be at this school?
Explain that for convenience all places in Britain use

London (Greenwich) time and not sun time. The difference in sun time is allowed for in lighting-up tables for cyclists.

26. Longitude II

By collaboration with another school some distance to the east or west the number of degrees of longitude between the two can be measured.

Using a long vertical pole, find as nearly as possible the exact moment when the shadow is shortest. It is best, perhaps, to mark the position of the end of the shadow every minute for about ten minutes before and after the correct time. It is essential to use exact Greenwich time.

Suppose at one school the time is 11.46, and at the other 12.3. Then, since there are seventeen minutes between the respective moments of local noon, the distance between the places of observation must be $17° \div 4 = 4 \frac{1}{4}°$

Teachers should note that even at Greenwich, sun noon does not coincide with clock noon except on 15th April, 15th June, 30th August, and 25th December. If the experiment is done about these dates, the interval between local noon and Greenwich noon will give the longitude of the school.

From the above it follows that a sundial can only be correct on the dates mentioned.

27. The Planets

The only planets that are ever conspicuous objects in our sky are Venus, Mars, Jupiter, and Saturn. A lesson on any one of these which can be seen may well arouse emotions of awe and admiration.

Details can be obtained from any good work on astronomy.

Venus, owing to the fact that its orbit is so near the sun, is always seen towards the west just after sunset or towards the east just before sunrise. It is therefore called either the Morning Star or the Evening Star. Owing also to its position it passes through phases like the moon, changing from crescent to "full". A telescope, however, is needed to perceive this.

Mars is generally recognizable by its red tint. Mention may be made that the ancients named it after the God of War, and the interesting conjecture that its red colouring may be due to the vegetation on its surface should be referred to.

The so-called "canals" may be described possibly as narrow belts of cultivated land near the banks of irrigation canals.

Jupiter, the largest planet, is still probably hot, and what we see is most likely merely the layer of clouds enveloping its surface.

Saturn has ten moons, and its ring probably consists of thousands of other tiny ones (it is transparent).

The sky, viewed from the surface of Saturn, must present a striking and beautiful appearance.

Plain as the glistening planets shine
When winds have cleaned the skies.

                                        Stevenson

Now the bright morning-star, day's harbinger,
Comes dancing from the east, and leads with her
The flowery May.

                                        Milton.

Gem of the crimson coloured Even,
Companion of retiring day,
Why at the closing gates of heaven,
Beloved star, dost thou delay.

                                        Campbell.

SECTION II
WEATHER OBSERVATION

To the child this is often one of the most interesting branches of Geography. He knows that the weather affects his own activities —the rain keeps him indoors, frost yields him slides, and so on. His interest, therefore, has a very practical foundation.

This initial interest is very valuable to the teacher, for the subject has two sides, each with its own educational value.

1. It is a valuable training in accurate observation.

2. It gives material for a proper understanding of climate.

The fact cannot be too strongly insisted upon that climate is simply *average weather*. Our own weather conditions are so diversified that we get samples, as it were, of the climate of nearly every region of the globe. We can obtain a clear idea of the fogs of Newfoundland, the storms of Cape Horn, the torrid heat of Bengal, &c, by comparison of our own weather periods of a few days with their conditions for weeks or months.

28. Heat and Cold

Even the youngest school children may be taught to observe some facts about weather, e.g. that sunny days are warm, that the sunshine is warmer towards the middle of the day than early or late, that south and west winds are generally warmer than north and east, &c.

On a sunny, breezy day let the class march round the

playground in single file so that part of the way is in the sun, part in the shade, part in the wind, and part sheltered.

1. Where was it warmest and where coldest?

2. Where does the heat come from, outdoors? [From the sun.]

3. Does a wind make us warmer or cooler? [Cooler.] Mention the use of fans, the punkah in hot countries.

4. How can we keep warm in cold weather? [1. By wrapping up our bodies. 2. By exercise.]

5. Why does walking against a strong wind make us warm? [We have to work so hard.]

Let some children dip their hands into water and then wave them about.

6. What happens to the water? [It dries up.]

7. Do you feel anything? [Our hands are colder.]

If possible, try it again with warm water. When water is drying up it causes cold.

8. Why is it bad to get one's clothing very wet? [As it dries it makes us cold, and we "catch cold".]

Describe the sun to the children as a great ball of fire, very hot, very large, and a very long way off. When it has warmed the air we feel warm even though the sun is not shining. A south wind is generally warm, because it comes from places where the sunshine is very hot. A north wind comes from regions of ice and snow, and is cold.

> When icicles hang by the wall,
> And Dick the shepherd blows his nail,
> And Tom bears logs into the hall,
> And milk comes frozen home in pail.
>
> Shakespeare.

29. The Thermometer I

Take a flask and a cork furnished with a glass tube. The bottom of the tube should be flush with the bottom of the cork. Fill the flask nearly to the brim with water, and push the cork in so that some water rises up the tube and no air is left in the flask. The water may be coloured with a little ink.

Tie a thread round the tube at the level of the water. Take it outside, carrying it by holding the cork, so as not to warm the water with the hand. Point out the behaviour of the water-level.

1. Is it warmer or colder outside than in the class room? [Colder.]

2. What has happened to the water? [It is now lower.]

Tell the children that liquids need more room when they are warmed, and less when they are cooled. When the water is warmed, and needs more room, it pushes its way up the tube, dropping down again when it is cooled and needs less room.

If possible, put the flask in the sun to show the rise; put it in a bowl of cold water; put the warm hand on it &c.

3. Why is this flask not good for work on a cold winter day? [The water might freeze.]

Tell the class that a liquid called mercury will not freeze, except, perhaps, in very cold countries, and it is better in other ways. Spirit of wine is also used. Show a real thermometer and compare the bulb, and the stem with the glass tube.

4. How is the thermometer different from our flask? [It is smaller.]

5. Why is this better? [It is handier, and the water gets warm or cold more quickly.]

30. The Thermometer II

Let the class make a drawing of an ordinary thermometer. Pay particular attention to marks 32°, 60°, and 80°. Tell them that 32° is the temperature when water begins to freeze, 60° is a comfortable temperature for a room, and 80° is too hot, though often the summer temperature rises above this. Show the effect of breathing upon or touching the bulb. Let temperature be taken at 9 a.m., noon, and 4 p.m. in the following places: the shade on the north side of a wall, the shade on the south side, and the sun. The thermometer must be left for about five minutes in each place before the temperature is read.

1. Why is this? [To give the liquid time to get warm or cold.]

2. Why must the thermometer not be breathed on or touched? [This would warm it.]

3. What is generally the hottest part of the day? [The afternoon.]

4. Why? [The sun has been up for a long time.]

5. What is probably the coldest time of night? [A little before sunrise.]

6. Why? [The sun has been away for a long time.]

Time.	Temperature.	Remarks.
9	51°	
9.30	51°	
10	51½°	
10.30	52°	
11	51°	The sky has clouded over.
&c.	&c.	&c.

Hang the thermometer outside in the shade and let it be read every half-hour during one day. Let a different boy go and read it each time. Record the observations as follows:

31. The Maximum and Minimum Thermometers

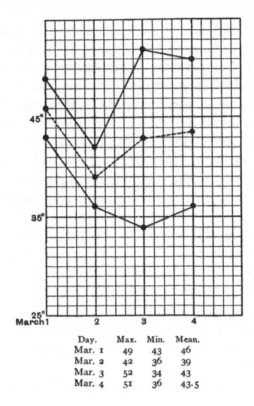

Day.	Max.	Min.	Mean.
Mar. 1	49	43	46
Mar. 2	42	36	39
Mar. 3	52	34	43
Mar. 4	51	36	43.5

The best form to use is Rutherford's. If possible, the two thermometers should be separate.

In the maximum thermometer there is a short iron wire to act as an index. When the temperature rises, the mercury expands and the index is pushed forward. If the mercury contracts, the index remains stationary. Consequently the end of the index *nearer the mercury* indicates the highest temperature reached. The index is set by drawing it to the mercury by means of a magnet, or by tilting the instrument till the index falls to its required position.

The minimum thermometer generally contains alcohol. The index is placed in the liquid, which drags the index with it as it contracts, but readily flows past the index when expanding. The lowest temperature is indicated by the end of the index *nearer the surface of the liquid*, as before. The instrument is set by tilting it.

The thermometers should be placed in a shaded position outside, at least 4 ft. from the ground. They should be read every day at a certain time, say 9 a.m., and then set ready for the next twenty-four hours.

For a long period, a year if possible, a table of records should be made and a graph drawn as shown. The maximum temperature graph should be in red ink. A continuous graph for a year (15 ft. long on scale given) on the wall of a classroom would be very valuable for reference, and would show the seasonal changes.

32. Rain—I

A few observations should be made during a wet day.

1. Is it dull or bright to-day? [Dull.]

Explain how the clouds shut off the sunlight.

Compare clouds to the vapour coming from a boiling kettle. Also refer to vapour condensing on a window pane.

2. If we went to a great height in an aeroplane, what should we see? [The sun shining above the clouds.]

Explain that drops of rain come from the clouds.

When there are many drops we say it is raining heavily.

Describe the rain in some tropical countries, where it seems as if a tank in the sky had been opened and the rain comes down in a solid sheet of water.

Tell also about places where it has never been known to rain.

3. Would you like to live there?

The answer will probably be "yes". A talk about the need

of water for plants and the consequent desert nature of rain-less areas may follow.

Other observations may be made, forming a starting-point for instructive talk, e.g.:

4. Are the drops always the same size? Very tiny drops are called mist or drizzle.

5. Do the drops of rain fall straight down?

6. What else falls from the clouds instead of rain? Tell about the astonishment of natives of hot countries when they see snow for the first time.

> And the angel spirit of rain laughed out
> Loud in the night.
> Loud as the maddened river raves in the cloven glen.
> Angel of rain! you laughed and leaped on the roofs of men.
>   Stevenson.

33. Rain—II

On a wet day put outside a large shallow tin such as a baking-tin, and also a tin of smaller area. Show the class how much water they contain after, say, six hours of rain. Put them on a level table and measure the depth of water.

1. What happens to most of the rain that falls on your garden? [It sinks in.]

2. Why is this a good thing? [It waters the roots of the plants.]

3. Does rain sink in the playground? [No, or very little].

4. What does happen to it? [It runs off.]

5. Where does it then go to? [The drains.]

Explain what these are, and how the rain reaches the local river and at last the sea.

6. Did any of the water that fell in our tins run off or sink

in? [No.]

7. Which tin has most water in it? [The larger one.]

8. Which has most depth of water in it? [Both the same.]
Explain that we can measure the rainfall by such a tin.

9. Rain disappears from the pavements sometimes without running off. How? [It dries up.]

10. When we want to dry wet things quickly what do we do? [Hold before the fire.]

A little talk about "drying up" might follow.

11. Is any water lost by drying up from our tin? [Yes.]

Show the school rain-gauge, and explain that the funnel is to prevent "drying up" as much as possible.

It will probably be best to speak as if the water was caught in the outer or large tin, and no reference to the glass measure or to "inches" need be made with such young children.

34. The Rain-gauge

Take a glass funnel and a jam-jar or a similar vessel of the same diameter as the funnel. Fitted up as shown in fig. 1, this will form a rough gauge accurate enough for most purposes.

A graduated, flat-bottomed cylinder will also be needed. Its diameter should only be about one-third that of the jar. Paste a vertical strip of paper, AB, on the side of it. Pour water in the jam-jar till its depth is just 1 in. (measure by a rule held vertically inside the jar). Transfer this water to the narrow cylinder—of course it will rise much more than 1 in. Draw a line on AB at the surface of the water and mark it 1 in. Subdivide this height into tenths and hundredths of an inch.

The children should be encouraged to make one for themselves at home for use in their own garden or yard. Call attention to the fact that the graduated cylinder is useless except for the particular gauge for which it is made.

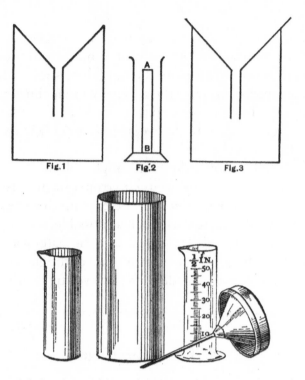

Fig.1   Fig.2   Fig.3

Should the funnel be of greater diameter than the jar, as shown in fig. 3, the above method will not do. The height of the 1-in. mark on the cylinder must be found by calculation.

Let A, R, D, be the area, radius, and diameter of the funnel.
Let a, r, d, be the area, radius, and diameter of the cylinder.
Let h be height of water when poured into cylinder.
Volume of water in jar = volume of water in cylinder.

$$1 \text{ in.} \times A = h \times a.$$
$$h = \frac{A}{a} = \frac{\pi R^2}{\pi r^2} = \frac{R^2}{r^2} = \left(\frac{D}{d}\right)^2.$$

Therefore the diameter of the funnel must be divided by the diameter of the cylinder, and the quotient squared will give the number of inches that the 1-in. mark must be drawn above the bottom of the cylinder. The above, of course, must be greatly simplified for young children.

35. Rainfall

On a wet day the children may be asked to estimate how long it will take for, say, ¼ in. of rain to fall. The length of time they give will probably be far too short. Put out a rain-gauge and find what falls in a measured time, say an hour or two.

In an arithmetic lesson the weight of water per acre that corresponds to 1 in. of rain may be found.

1 ac. = 4840 sq. yd.

Volume of water if it lay 1 in. deep over this area =4840 × 9 × 144 × 1 cu. in. = 3630 cu. ft.

1 cu. ft. of water weighs about 62 ½ lb.

Weight of this volume of water = 3630 × 62 ½ lb. = 101.3 tons approx.

We often say that 1/100 in. of rain = a ton per acre (approx.)

The huge amount of water that falls even during a slight shower may now be appreciated.

Figures such as the following may now be discussed:—

Place.	Mean Annual Rainfall.
Near Keswick.	177 in.
Spurn Head.	19 in.
Cherripungi (Assam).	439 in.
Atacama (Chile), &c.	0 in.

The heaviest fall recorded in one day in Britain occurred near Keswick —8.03, in. on 12th November, 1897. During a typhoon in the Philippine Isles 35 in. fell in one day. The heaviest rate recorded in Britain is ⅓ in. in two minutes. But even an inch in a day is regarded as heavy rain in our country.

The above figures must be compared and contrasted with those obtained at the school.

36. Wind

A short talk about air should bring out (1) that we can feel air though we cannot see it, (2) that moving air is wind, (3) that wind can be heard and can sometimes move quite heavy objects.

1. What do we call a very light wind? [A breeze.]

2. What do we call a very strong wind? [A storm or gale.]

3. What have you seen the wind move during a gale?

4. Where do we feel the wind most? [At the top of a hill.]

5. Why is this? [There is nothing to shelter us.]

Explain that we name a wind according to where it blows from.

6. Which winds are likely to bring rain? [Those from the sea.]

Speak of the south-west wind, usually warm and wet. Refer to the hat call a "sou'wester".

Speak of Spain, Italy, and other countries to the south, Russia, with its hot summers and cold winters, to the east, &c., and elicit the reasons for the nature of, e.g., a south wind.

Call attention to the waves formed on the largest sheet of water available. With children in inland districts an account of the waves of the sea should follow.

Welcome, black North-easter!
O'er the German foam ;
O'er the Danish moorlands,
From thy frozen home.

Kingsley.

37. Clouds

On several days the clouds should be noticed, so that the following questions may be answered: —
1. Was the whole sky covered or only part?
2. Which covered more space, the blue sky or the clouds?
3. Were the clouds white, grey, or nearly black?
4. Was rain falling from them or not?
5. Were they moving, and, if so, which way?
6. Were they moving slowly or quickly?
7. Did they look higher than usual or lower?
8. Do the clouds ever touch the hills?
9. Were they coloured when the sun was setting?
10. If so, what colours did you notice?

If the clouds were going westward there must have been an east wind up there. The wind is often stronger high up in the air, and sometimes does not blow quite in the same direction as on the ground.

Give the following rhyme, explain it, and let it be memorized:—

"Evening red and morning grey,
This the sign of a very fine day;
Evening grey and morning red,
Put on your hat or you wet your head".

When possible, the four chief varieties of cloud should be pointed out—cirrus, cumulus, stratus, and nimbus. Of course, with young children these names should not be given, but they may be referred to as feather clouds, woolly clouds, belt or line clouds, and rain clouds respectively.

I bring fresh showers for the thirsting flowers,
From the seas and the streams;
I bear light shade for the leaves when laid
In their noonday dreams.

Shelley.

See the heavy clouds low falling,
And bright Hesperus down calling
The dead Night from underground;
At whose rising, mists unsound, Damps and vapours, fly apace.

J. Fletcher.

38. Wind Directions

Take the class into the playground and show them how the direction of the wind may be found. Let them clearly understand that we name a wind according to where it comes *from*.

1. Why cannot we judge the direction by feeling it on our faces? [Buildings alter its direction.] Where could we use this

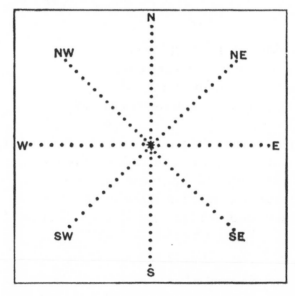

method ? [In an open space.]

2. Where should a wind-vane be put? [Above any buildings or trees.]

3. Where does the head of the weathercock face? [Towards where the wind comes from.]

4. What can we see at a height which shows wind directions? [Smoke and clouds.]

5. Which is the better? [Clouds are less affected by buildings, &c.]

Tell the class that sometimes the wind at a great height does not move quite in the same way as near the ground.

6. Let a child moisten his finger and hold it up. Which side feels colder? [That towards the wind.] Why? [That side dries more quickly.]

Explain that clouds nearly overhead are the best to watch; it is hard to judge the direction of those near the horizon.

Take a piece of cardboard and put dots on it ¼ in. apart as shown.

Provide eight pins and let one be moved one division along its correct line each day that the wind is in that direction. Start from the centre. If this is done for a month it will show the comparative frequency of each wind-direction during that period.

39. Dew, Fog, and Mist

On a clear still evening lay some stones, pieces of wood, slate, &c., outside. Next morning they should be examined soon after sunrise, together with grass, gravel, &c. Try this on other evenings. Set the following questions:—

1. Which substances have most dew and which least?

2. Are the upper surfaces or the under (when exposed) most bedewed ? [Under.]

3. Is it really correct then to talk of dew "falling"? [No.] Explain how the moisture of the air is deposited on a cold surface.

4. Does dew form well under trees or bushes, and why? [No, the escape of heat is checked.]

5. Does it form well on cloudy nights. [No, same reason.]

6. Does it form well on windy nights? [No, the cool air near the ground becomes mixed with warmer air from above.]

Hoar-frost is caused by the dew freezing as it forms.

Explain that fog and mist are really cloud that has descended to the ground. We say that we can see a cloud touching a high hill, but if we ascend the hill and enter the cloud we speak of it as mist. When the children go out in a fog let them observe the tiny drops of water forming, e.g. on their clothing. A piece of fine muslin may be tied round the mouth and nose as a respirator. Let them notice the large amount of dirt that has settled on it where the breath has passed through. This is particularly noticeable in towns. Hence speak of the importance of breathing through the nose in order to filter the air, and not through the mouth.

An experiment may be devised to show that foggy air conducts sound particularly well. For example, a pencil may be dropped from a measured height and the distance noted at which a child can just hear the sound. In clear air he will not hear it at this distance. Refer to fog signals, sirens in ships and lighthouses, &c.

—As silent as the dew comes
From the empty air appearing,
Into empty air returning,
Taking shape when earth it touches;
But invisible to all men
In its coming and its going.

                                        Longfellow.

40. The Barometer

Obtain a tin with a small opening and put a little water in it. Boil the water, and, when steam is issuing freely, cork the opening very tightly, removing the source of heat at the same time. As the steam condenses, the pressure inside diminishes and the tin crumples up.

Another well-known experiment is to fill a tumbler to the brim with water, put a piece of cardboard over it, and invert the tumbler. The cardboard will now remain in position if the hand be withdrawn.

These and other indoor experiments will prove that the air exercises pressure. An aneroid barometer should now be shown. It consists of one or more shallow metal boxes with movable lids. When the outside pressure increases, the lid is pushed in. The movement is communicated to an index finger which moves round. The boxes are exhausted of air, or differences of temperature would expand or contract the air and cause movements of the index.

The mercury is useless for outdoor work, but reference

must be made to the fact that we measure air-pressure by the
number of inches of mercury it will support. In this country,
at sea-level, the pressure is practically always between 28 in.
and 31 in.

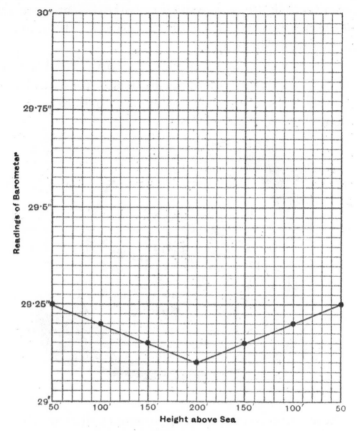

Prepare a piece of squared paper as shown. Take the aneroid along a hilly road to observe how the pressure falls with
increase of altitude.

Take similar readings on another paper, from the barometer
in a fixed position to show how the pressure varies according
to the state of the atmosphere.

41. Weather Record I

For very little children the best form will be pictorial in character. Some adaptation of the following may be found useful:—

Prepare seven rectangles, say 2 in. by 1 in., one for each day in the week. Let the child colour a rectangle according to his idea of the appearance of the sky that day. A little yellow circle for the sun would indicate a sunny day. The rest of the rectangle will be coloured to represent the sky, e.g. dark for clouds, blue for clear, slanting broken lines for rain, &c.

To show temperature, such words as " hot", " cold", "warm", "very cold" may be written by the side of each oblong.

Each day the best one may be fixed on the classroom wall, so that a record for a period is obtained. Again, a large circular chart may be prepared and divided into sectors, named and coloured according to the above suggestion. A movable pointer may be fixed in the middle, and turned each day to the sector most nearly representing the weather conditions.

On days when the weather is rather exceptional, a talk should be given about other climates, e.g. on a snowy day reference may be made to people who have never seen snow, other people who live in snow houses and scarcely know what rain is, &c.

42. Weather Record —II

Let the children rule a page of their notebooks as follows: —

Date.	9 a.m.	1 p.m.	5 p.m.	9 p.m.

Let them learn the following scale. It is known as the Beaufort scale, after Admiral Beaufort who designed it.

b. Blue sky.	p. Passing showers.
c. Detached clouds.	q. Squally.
d. Drizzling rain.	r. Rain.
f. Fog.	s. Snow.
g. Dark, gloomy.	t. Thunder.
h. Hail.	u. Ugly, threatening.
l. Lightning.	v. Visibility, unusually
m. Misty.	clear air.
o. Overcast.	w. Dew.

Explain the exact meaning of each term and let the class keep a record for a fortnight or so, using these terms only. For example, b. c. p. would describe a period of blue sky with detached clouds and occasional showers.

At the same time the thermometer should be read (Lesson 30) and the direction of the wind noted (Lesson 38).The force of the wind should be estimated by reference to the following scale:—

Description.	Speed.	Effects.	Sign on a Map.
Calm	o	—	☉
Light	2–5 m.p.h.	Moving leaves ... }	—
Moderate ...	7–10 „	„ branches ... }	
Brisk	18–20 „	Swaying branches ... }	⟶
High	27–30 „	„ trees ... }	
Gale	45–50 „	Breaking branches	⟼
Severe gale ...	above 50 „	Damage to buildings	⟼

43. Weather Record III

Older children should keep a systematic record of readings of the various instruments. They should compare this with the record at the nearest meteorological station, or with those published in the daily papers.

A graph of the temperature should be made as shown

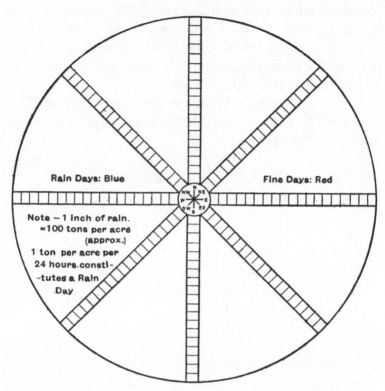

Rain Days: Blue

Fine Days: Red

Note — 1 inch of rain.
≈100 tons per acre
(approx.)
1 ton per acre per
24 hours.consti-
-tutes a Rain
Day.

in Lesson 31. A similar graph should show pressure and rainfall, and another the altitude of the sun. If such graphs are displayed together on the wall of a classroom, it will be easy to point out cause and effect, e.g. the increase of sun's altitude resulting in increase of average temperature, a fall in barometer causing rain, &c.

A wind and rain chart may be used. It will indicate the wind directions for a given period, and also show which winds were rainy and which were dry.

Suppose that on 1st March there was an east wind. Then a pencil cross would be made on the first section to the right of the centre. On 2nd March the rain-gauge would be observed, and if more than 1/100 in. had fallen (1 ton per acre) this section would be coloured blue. If less than this amount, it would be coloured red. Of course this is an arbitrary limit—in a wet district less than $\frac{1}{50}$ in. may be considered as fine.

At the end of a month there will be 31 sections coloured, and it will be seen at a glance which were the prevailing winds, which the dry, and which the wet.

The "Scarborough" Weather - observation Chart (Philip Harris & Co., 1s. each) gives further information, and is ruled ready for all purposes.

Season of mists and mellow fruitfulness!
Close bosom-friend of the maturing sun.

Keats.

But howling Winter fled afar
To hills that prop the polar star;
And loves on deer-borne car to ride
With barren darkness at his side.

Campbell.

The blasts of Autumn drive the winged seeds
Over the earth,—next come the snows, and rain,
And frosts, and storms, which dreary Winter leads
Out of his Scythian cave, a savage train ;
Behold ! Spring sweeps over the world again,
Shedding soft dews from her sethereal wings:
Flowers on the mountains, fruits o'er the plain,
And music on the waves and woods she flings.

Shelley.

When first the fiery-mantled Sun
His heavenly race began to run,
Round the earth and ocean blue
His children four, the seasons, flew.

Campbell.

44. Weather Forecasting

Simple rules should now be deduced from the records, e.g. that a falling barometer foretells wind or rain, a continued fall means a long period of bad weather; an east wind often leads to dry weather, cold in winter and spring, hot in summer.

Let it be particularly noted that the words written on the dial of a barometer—fair, rain, &c.—are not to be relied upon. If the index is pointing to rain, for example, but is slowly and steadily rising, the weather is likely to be fine.

Encourage the children to learn weather signs from country people, especially sailors and farmers. Let them test the truth or otherwise of the following beliefs:—

1. That swallows fly high before fine weather, and low before rain.

2. That when distant hills, &c., appear near, rain will follow.

3. That weather changes with the tide (near the coast).

4. That weather that commences at new moon will last long.

5. That when smoke does not rise well, when it emerges from a factory chimney, rain is likely.

Hang up out-of-doors (1) a piece of broad, flat sea weed, (2) a piece of catgut (fiddle-string), and note how they change according to the dampness of the air.

SECTION III
PLANS AND MAPS

A great deal of work in copying maps is done in most schools. It may be a very valuable exercise, but its value will depend largely on the child's knowledge of the meaning of a map. Such knowledge can only be obtained by constant comparison of the conventional marks on a map with the realities which such marks represent.

Some actual map-making should precede map-copying. In point of accuracy such efforts may be very poor, but even then the work of construction will have called for observation and thought of the greatest possible value.

When teachers find it inadvisable to take the whole of their class outdoors, half of them may remain in school to copy maps or do the written work in connection with previous exercises.

A warning may be given about so called "practical" exercises that will give no correct results. The curvature of the earth, for example, cannot be proved by experiments in a playground. In a furlong the allowance for the curvature is only ⅛ in., so it is obvious that excessively slight inequalities in the surface will quite neutralize the effect we wish to observe.

45. The Compass—I

Show the class a large compass, or, better still, lend a small one to each group of four or five (see Lesson 46 for a method of making a rough compass if one can not be procured). Let them notice that however the box is turned the needle always points in the same direction.

Tell them that one end always points north, but no one understands why it does. Show that it points in the same direction (nearly) as the noon shadows.

Practise in the following kind of question:—

1. Face north. Point west.

2. Face south. Point east.

3. What place (village, building, hill, &c.) is north of us?

4. Face towards X (some well-known place). What direction is it?

5. If you went to X, in what direction would this school be?

Cut a circular disk of cardboard, the same diameter as the needle, and mark on it the N., E., S., and W. points. Fix it on the top of the needle, and show how it swings with the needle, thus giving the other directions correctly.

46. The Compass—II

Take a steel knitting-needle. Stroke the needle with one end of the magnet, always stroking in one direction. The needle will itself become a magnet, and will be able to lift up a few iron filings or a pin.

Tie one end of a piece of thread round the middle of the needle and hang it up, well away from any iron object. After swinging for a time the needle will come to rest.

Which way does it point? [About N. and S.]

Show a large compass. A similar arrangement is used there, only the needle swings on a sharp point.

Teach the meanings of S.E., S.W., N.E., N.W.

Give practice with these terms.

If time permits, take the compass to the north-and-south

line drawn in the playground, and show that the compass does not point exactly north but a little west of north, and, of course, east of south.

Show that any piece of iron or steel will attract the compass and pull it from its proper position, but other substances, such as brass or wood, have no effect on it. Show that the compass does not point N. if it is too near a gas-lamp, iron railings, &c.

47. The Compass—III

Put a compass needle on the north-and-south line which has been marked in the playground (see Lesson 2). Let a boy go several yards away with a stick, and tell him to move to the right or left till the stick (held vertically) is in a line with the needle. By means of a piece of string, draw

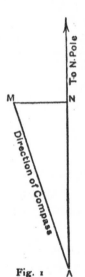

Fig. 1

a chalk line from the stick to the position of the needle. This line is called a *magnetic* meridian, and will make a small angle with the true north-and-south line or *geographical* meridian.

The angle between the two is called the magnetic declination for that place. To measure it, mark off along the geographical meridian a distance of 15 ft. from the junction of the two lines. Draw from this mark N at right angles till you come to the magnetic meridian.

Measure this cross line NM. Now draw on paper, using a scale

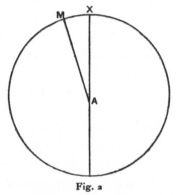

Fig. 2

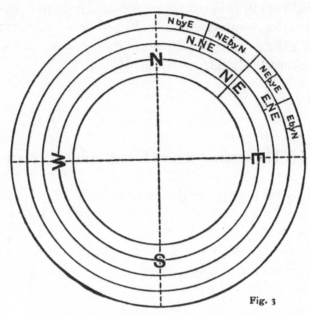

Fig. 3

of 1 in. to 1 yd., i.e. your distances will be 1/36 of those in the playground. Now measure the angle NAM with a protractor, or else compare it with the angle of 30° on a set square, and judge what it is.

Another way would be to draw a circle (with chalk at the end of the string) 15 ft. radius and centre at the compass. With string find how many times MX divides into the whole circle. Suppose it is 1/24 of the whole. Then the angle is 1/24 of 360° = 15°.

Copy and complete the diagram. The five circles should

TABLE OF COMPASS POINTS

Compass Point.	Angle in Degrees East of North.
N.	0°
N. by E.	11¼°
N.N.E.	22½°
&c.	&c.

have radii of 1 in., 1 ¼ in., 1 ½ in., 1 ¾ in., and 2 in. How many points of the compass are there altogether? Construct a table showing the position in *degrees* of each of the points of the compass, reckoning from N. eastwards.

48. Determination of Geographical North

In this lesson the three different methods should be correlated, (1) by the sun, (2) by the Pole Star, (3) by the compass.

A north-to-south line has already been drawn (Lesson 2). From a point on it two other lines should be drawn north, using methods (2) and (3) above. These lines should of course coincide. Probably they will not exactly do so, owing to (a) experimental error, (b) the fact that the Pole Star does move, though very little, (c) the difficulty of determining exact local noon.

For Method 2, let a boy put up a pole A. Let another boy go to a distance with a very long pole held vertically, and let the boy at A direct him to right or left till from A the two poles are in a line with the Pole Star. The position B can then be marked, and the line drawn next day.

For Method 3 it will be necessary to draw from A the compass direction or the magnetic meridian as in Lesson 47, fig. 1. The magnetic declination at the school must also be known (see Appendix V).

Draw AM a magnetic meridian. Mark off the angle MAD

equal to the magnetic declination. A very large protractor should be used, such as the one on the clinometer (see Lesson 70).

The lines AB, AC, AD are then all approximations to a true geographical meridian.

49. Introduction to Plans

Measure a rectangular playground or large room by means of a stick just a yard long. Fractions of a yard may be neglected.

A plan may then be drawn on a very simple scale, e.g. 1 in. to 1 yd.

Point out that the general shape of the plan is like that of the original. It will help if some different shape be drawn on the board, e.g. a very long narrow oblong to represent nearly a square yard. The children will notice the error, and will then see more clearly that the scale drawing is correct.

Make a plan on the board, show how many feet long each side is, and ask what it represents on a scale, e.g., of 1 yd. to 1 ft. or 1 ml. to 1 ft.

50. Introduction to Maps—I

Make an extremely simplified copy of part of the 6-in. map (see Appendix II) of the school district. If possible cyclostyle it, so that each child can have a copy.

Take the class to the place and let them compare the map with the original. There should be examples of such signs as shown on the following page:—

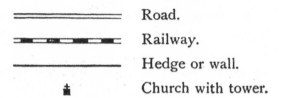

The meanings of the conventional signs should be thoroughly learnt before going out.

Take a compass with you and let the north be found. Mark a place on the map, and let observations be made from there. Point out that the top of the map represents the north. Let the children turn that way with their maps, and notice that objects marked to the left on the map are really to the left.

Walk along one of the roads, and let the children say from their maps what they will soon come to, e.g. a side road, church, &c. Let them estimate how long it will take to reach it.

Some estimation may be made of distances, e.g. as clear an idea as possible of what is meant by a mile should be given.

Several lessons on the above lines should be given, so as to accustom the children gradually to understand the meaning of a map.

51. Introduction to Maps—II

Let the class go out to a suitable place, each provided with a pencil and a blank paper mounted on a sheet of stiff cardboard.

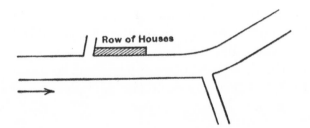

Without making measurements, let them draw a plan of the road, any buildings, side roads, &c., somewhat like the one shown.

Tell them to imagine a tiny insect walking on the map. Going in the direction of the arrow it will come to a lane going to the left, just as we do in reality, then to a row of houses built up to this lane, and so on.

When they have got the idea, the map may be revised according to distances, which should be measured by steps. For example, 10 paces might be represented by ¼ in.

The conventional signs used on maps must be taught, not too many at once. See the characteristic sheets for 1-in., 6-in., and 25-in. maps (Appendix II).

The children should then be encouraged to prepare and show a plan of the neighbourhood of their own home.

52. The Chain—I

Take a ball of string to the playground, unwind it and pull it tight. Let a boy with a ruler measure accurately 66 ft. and cut it off at that length. This is the length of one chain.

Prepare a number of cardboard tabs as illustrated. There should be one like No. 5, and two of each of the others.

Double the string to find its centre, and fasten No. 5 there by coloured thread. From each end of the string measure 6 ft. 7 ⅕ in., and at each place put a No. 1 tab. Another 6 ft. 7 ⅕ in. nearer the centre (best got by folding the string) put a No. 2 tab, and so on with Nos. 3 and 4. If you have measured correctly, the distance from the two No. 4 tabs to No. 5 will

1   2   3   4   5

also be 6 ft. 7 ⅕ in., and the chain will be divided into 10 equal parts. Each part is 10 links long.

With the chain stretched out let a number of boys mark the links with coloured thread. The intervals between them should be about 7 ¹¹⁄₁₂ in.

Teach the table:—

100 links = 1 chain.	1 link = 8 in. (nearly).
10 chains = 1 fur.	1 chain = 66 ft. = 22 yd.
80 chains = 1 ml.	

Mention some place about a mile away. The children should have some idea of what is meant by a mile.

53. The Chain—II

Prepare ten pieces of wood each about a foot long and pointed at one end so that they will stick in the ground. These are called arrows.

Take part of the class to a road, preferably one with quarter-milestones, and let them measure a long distance. Only two boys must have the chain at one time. The leader takes the ten arrows and one end of the chain. When he has pulled it tight, he puts one arrow in the ground and moves on again. The follower sees that his end is just against the arrow, and then picks the arrow up when they move on. The number of arrows in the hand of the follower shows the number of chains traversed.

At the same time a few boys might measure the distance by pacing. Each should first have paced along the chain. By dividing 66 ft. by the number of steps he will find the average length of his stride. Multiplying this distance by the number of steps taken in going along the road will give the length of the journey. Remember that people naturally take a shorter stride when going uphill and a longer one in going downhill. If the road to be measured is hilly, it is therefore more accurate to pace it in both directions and take the average of the two results.

54. The Chain—III

Take the chain to the playground or a small field and lay it along the greatest diagonal, AC. With another piece of string measure the distances from the corners F, E, D, B, to the line AC. All the measurements must be at right angles to the chain. Prepare a "field book" as shown. Rule your page into three columns and start at the bottom of the page, working upwards. In the centre column put the distances of G, H, K, L, and C from A, all measurements made in links. AG = 8, AH = 17, AK = 20, AL = 25, AC = 32 links. In the left column put the measurements made to the left, and in the

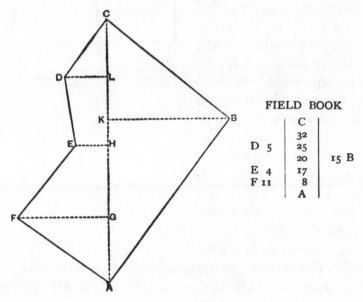

FIELD BOOK

		C	
		32	
D	5	25	
		20	15 B
E	4	17	
F	11	8	
		A	

right column those made to the right. As we start from A and go to C the first two measurements are to the left, the next (KB) to the right, and the last to the left.

Now take a piece of paper ruled in ¹⁄₁₀ in. squares. Draw a vertical line to represent AC on a scale of ¹⁄₁₀ in. to 1 link or 10 in. to 1 chain. At the correct distances from A mark off G, H, K, and L. Measure from these points to right or left as the case may be and you will have the points F, E, B, and D. Join these as in the drawing and you will have a plan, correct to scale, of the piece of ground.

55. Chain Survey Plan

For more accurate work with the chain a "cross head" or "off-set finder" will be needed. The lines FG, KB, &c., in Lesson 54 are called off-sets, and they should be exactly at right angles to AC.

In the woodwork lesson let five square pieces of wood be prepared. The thickness does not matter, but four of them must be exactly of the same size and rather less than a quarter the area of the fifth piece. The large piece might be 12 in. × 12 in. and the others 5 ¾ in. × 5 ¾ in.

Screw one of the four smaller pieces on top of the large one so that at the corner A the edges coincide. Do the same with the others at the corners

Plan

Elevation

B, C, and D. Then two grooves half an inch wide will be left, exactly at right angles to each other.

If the arrangement is held so that one groove is in the same direction as the chain, the other groove will help to fix the direction of the off-set. An upright support about 3 ft. 6 in. high should be provided.

Repeat the work of Lesson 54, using this instrument.

56. Chain Survey (Area)

Lesson 54 may now be carried further by finding the area of the piece of ground. Suppose that the side of each small square represents I link. Then a square will represent a square link.

Count the number of squares in the drawing. This can be done rapidly by dividing the drawing into rectangles and triangles, e.g. if FE be produced to meet CB, and a perpendicular be dropped on to it from D, the diagram will be divided into two rectangles and a triangle.

If a boundary line cuts through certain squares it is advisible to count all parts greater than halves as if they were whole ones, and to leave out altogether the parts less than halves.

Having counted the squares, we have the number of square links in the ground.

Teach the table :

$$10,000 \text{ sq. links} = 1 \text{ sq. chain}$$
$$10 \text{ sq. chains} = 1 \text{ ac.}$$

57. To Draw a Right-angle

It is often necessary to draw a line at right angles to another on a larger scale, where instruments used for paper work are useless. A builder, for example, needs to make one wall at right angles to another. The exercise is also valuable in its bearing on geometry. Let the class prepare three cords, 3, 4, and 5 units in length respectively. The unit may be a foot, a yard, ten links, or any other convenient length.

Suppose AB is a wall and it is desired to draw a line at right angles to it from A. With one cord measure AC, 4 units from A. Hold the 5-unit cord with one end at C and the other pulled in the direction D. Similarly, hold the 3-unit cord with one end at A. Where the ends of the two cords meet will be the point D, and DA will be at right angles to AB. As a check, mark E 3 units from A. With centre E measure 5 units, and with centre A measure 4 units. The point of intersection will be F, and the line FA should pass through D.

This will lead to an indoor lesson dealing with the connection between these lengths 3, 4, and 5, and to the fact that 5 squared is equal to the sum of 3 squared and 4 squared.

58. Measurement of Width of River—I

If no stream is at a convenient distance, mark out an imaginary one with chalk in the playground. Or on a country road the class can pretend that the road itself is covered with water.

Pick out some tree, stone, or other object A on the other bank. Just opposite to it fix a stick B vertically. Walk from here in the direction BC at right angles to AB until the angle ACB is 45°. This can be measured by holding a 45° set square horizontally up to the eye. When one edge points to A and

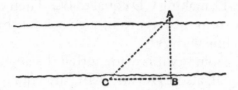

the other to B the angle is 45°. Now measure the distance BC either by pacing (see Lesson 53) or by the chain. This distance will equal the distance AB.

Some conception of the width of certain great rivers should be given.

59. Measurement of Width of River—II

This lesson will be similar to No. 58, but the method depends on the fact that two triangles are equal if they have one side and two angles of the one respectively equal to the corresponding parts of the other.

Fix upon A and B as before. Walk along the bank from B any convenient distance to C and there erect a vertical stick.

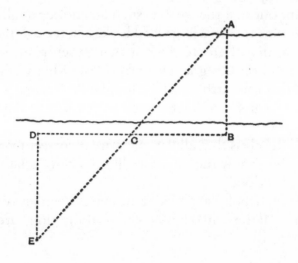

Continue to D, making CD equal to BC. Then walk at right angles away from the river to E, such a point that from it C appears in a line with A.

Then by geometry it is evident that the angles DCE and ACB are equal, also EDC and ABC are equal, both being right angles, and the side CD equals BC. Therefore the side ED equals AB.

Of course it is not necessary that the children should understand the above geometrical proof.

60. The Globe and Maps

In the playground take a long piece of string (say 10 yd.), fix one end of it, and attach the other to a piece of chalk. In this way draw on the ground a part of a very large circle. If the part drawn is only short, say 2 ft., it will look straight.

1. Is the line really straight? [No.]

2. How do you know that? [Because it is part of a circle.]

3. Why does it look straight? [Because the full circle would be so large.]

Point out that the earth is such an enormous globe that for the same reason a large field looks flat, though really it is part of a curve. Refer to the way that an aeroplane seems to be descending as it goes away from us. Other well-known proofs that the earth is a globe may be dealt with.

Show a globe, and let some comparison be made with flat maps.

Tell the class that all flat maps are incorrect in one way or another. Show the impossibility of laying a flat piece of paper on a globe.

4. Which looks larger from the map, Greenland or South America? [If the map is on Mercator's projection, Greenland looks larger.]

5. Which does the globe show really the larger? [South America is over twelve times the area of Greenland.]

Find on the globe the shortest way from Constantinople to New York by aeroplane. Compare with the map (Mercator's). Find what places you would pass in going from London to the South Pole. Compare with map [world in hemispheres]. In each case the flat map is wrong, because the earth is round and cannot be correctly drawn on a flat sheet of paper.

61. Lines of Longitude and Latitude

In the playground draw a north-and-south line.

1. Where would this line reach if carried far enough north? [To the North Pole.]

2. Where, if carried far enough south? [To the South Pole.] It is called a line of longitude. Show such lines marked on a globe, and some on a map of England.

Usually we only draw 360 such lines, starting from the one through London and taking them at equal distances all the way round the earth. Any line, however, such as the one we have drawn is a line of longitude, though that particular one may not be marked on a map.

Draw a line at right angles to the north-south line. This is a line of latitude.

3. Where will this reach if carried far enough? [Back again to the starting-place.]

Deal with this as you have with the line of longitude.

4. Are all lines of longitude the same length? [Yes.]

5. Are all lines of latitude the same length? [No.]

6. What shape are both sets of lines? [Circles.]

The term " great circle" may be used for the greatest circles that can be drawn on a globe. All lines of longitude are great circles but no lines of latitude except the equator. There is an infinite number of other great circles. Take a thread equal in length to one quarter the circumference of the school globe and tie a piece of chalk to one end. Show that, if the other end be fastened at the North or South Pole and a circle be drawn using the string as a radius, we can draw the equator. If a point on the equator is chosen as centre we draw a line of longitude. But if some other point is taken as centre we get another "great circle".

62. Size and Shape of the Earth

The diameter of the earth is a little under 8000 ml. Therefore on a scale of $\frac{1}{10}$ in. to 1 ml. it should be possible to represent it in the playground. 800 in. is just over a chain, so a chain will represent the diameter.

Hold the middle of the chain at a point, and fasten a piece of chalk at the end of the chain. In this way draw a circle with ½ chain radius.

Examine a short part of the curve, say 2 or 3 in., representing 20 or 30 ml. Point out that, though evidently part of a circle, it looks like a straight line. Hence the impossibility of seeing the curvature of the earth when we look at a plain or the sea.

Tell the class that the earth is not a perfect sphere, even if we neglect mountains, &c. The diameter through the poles is about 28 ml. shorter than one through the equator. Instead of our circle we should have drawn a curve with one diameter about 3 in. less than the other. Point out what a very small departure from a true circle this makes—one quite inappreciable by the eye.

Put marks on the circle to represent the poles. Draw the equator and insert the position of your town, e.g. if it is latitude 52 it will be $^{52}/_{90}$ of the way along the quadrant from the equator to the North Pole. Draw a horizon line there, and point out that a straight line to the horizon, if continued, goes away from the earth. To point to the North Pole one must point downwards through the earth.

63. The Plane Table—I

Take a drawing-board, pin a sheet of paper on it, and put the board on a high stool in the playground. Take care that the stool is not moved by accident. Call this pin position A, and insert pin (*a*) into the paper to represent this spot.

With the chain, measure a distance of 1 chain (or half this if the playground is small. Put up a pole B at the other end of the measured distance.

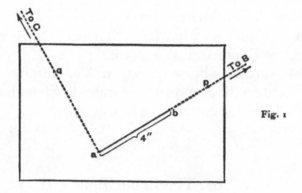

Fig. 1

Let a boy look along the line AB and put a pin (*p*) on the paper, so that A (*p*) and B are in a straight line. Join *ap* on the paper.

Choose a scale, say 4 in. long on the paper. This will represent AB on the ground.

Sight a post or corner of the playground C. and put in a pin *q* so that *aq* and C are in a straight line. Join *aq*. The board will now look like fig. 1.

Now take the board to B and put up the pole at A. The board must be put exactly parallel to its former position. This is best done by putting the pins a and p in again and looking back at the pole at A.

Then turn the board till the two pins are just in a line with the pole at A. The line ab on the paper will now again be in the same direction as A B on the ground.

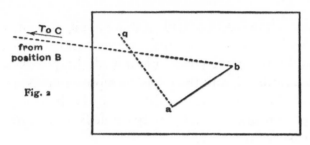

To C
from position B

Fig. 2

Next draw from *b* towards C just as you did from *a*. The board will now look like fig. 2. Where the lines from a and from b to C intersect is the point c.

Measure ac or bc on the paper. What distance does this represent on the ground by our scale? Test it with the chain; it should be correct within a foot or two.

64. The Plane Table— II

The plane table may now be used to construct a map of a small area. It is better to mount the table on a tripod (see Appendix I).

Before beginning the work the table should be turned so that the edge of the paper runs due north and south. This may be done by the trough compass (Appendix I). Place the edge of the compass-box on the edge of the paper, set the compass swinging by moving the lever, and then turn the board till the needle points to an angle equal to the magnetic declination. The edge of the box, and therefore of the paper, will then be geographical north and south.

Remember that accurate results cannot be obtained if the lines drawn from the two ends of the base line are nearly parallel.

A base line AB should only be used for such points as C, D, &c., but not for E and F. For the latter a new base line, e.g. BC should be used.

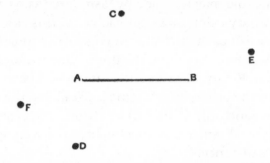

The length of the base line will depend, of course, on the area which it is desired to bring into the map. For a beginning it may be, say, 1 chain long, measured on the playground, and the scale may be 1 chain to 1 inch. Such points as the corners of the playground, the ends of the school, &c. may be found. Lines may then be joined representing walls, &c. Construction lines should be dotted.

65. The Plane Table—III

A much longer base line may now be measured on a straight piece of road or on an open space. The line should be as level as possible, but a slight gradient makes very little difference to the results.

The scale may be, say, 1 furlong (10 chains) to 1 inch, and the map will include objects a mile away or more.

The sight rule mentioned in Appendix I will give more accurate results than the arrangement of pins given in Lesson 63. Place a pin at the point of observation, on the paper, and bring the rule up to it. Then turn the rule, using the pin as a pivot, until the two vertical threads on the sighting-pieces are in a line with the object. The eye should be about a foot from the nearer thread.

A sight rule may be made by fastening two vertical pins in an ordinary ruler, one near each end. It is essential that the distance of each pin from the edge of the ruler should be exactly the same. Note that the line is drawn along the edge of the ruler. Really, of course, the line should be drawn from one pin to the other, but this line is parallel to the one actually drawn, and only an inch or less from it, an error which is extremely small, since the object sighted is at a comparatively enormous distance.

Some books give careful directions for levelling the plane table, but in practice a slight dip makes no difference to the results. A spirit-level may be used.

66. Setting an Ordnance Map

Take the 6-in. map, fastened on a drawing-board, to a position where a good view can be obtained. It may be put on a stool, or, better, on a tripod.

Insert a pin in the map at the place where the observations are made, and another on the representation of some prominent landmark. Turn the map till the two pins are in a line with the landmark. Several children should look to see that the line on the map really corresponds with the line across the country.

If this is true for one line it will be true for any other line. Keeping the map still, put a third pin on some other point on the map, and show that the line joining the pin will point to this second object in the distance.

Practise the class in judging distances, verifying from the map. The scale is nearly 300 yd. to the inch.

If time admits, a similar method may be employed with the 1-in. map.

Lay the compass needle on the map. As the top of the map is pointing to geographical, or true north, the compass will point to the left of this. The angle of declination may be found (i.e. the angle which the compass makes with true north).The trough compass (Appendix I) is best for this.

SECTION IV
HEIGHTS AND CONTOURS

The relief of the land is of great importance with relation to its effect on the course of rivers, roads, canals, and railways, and on climate. Consequently much importance is now attached to a study of contour lines.

Some excellent work can be done indoors, to explain the meaning of a contour line, by models and a tray of water. Further good work can follow with maps and sections. But it cannot be too strongly emphasized that contour lines refer to levels *of the ground*, and that therefore no real knowledge can be obtained without considering the real contours of the land as well as representations of them on paper or clay.

Some such work may, of course, be done without guidance, but usually the attention of the children when outdoors is so much taken up by other matters that even a good treatment of the subject purely by indoor lessons will lead to little knowledge of real contours.

67. Heights

Let the class stand within a few feet of a fairly high wall and let them look at the top row of stones, or at some other line at a height.

1. What have you to do to see something so high? [Put our heads back or look upwards.]

Take them to a distance of several yards and let them

look again.

2. Do you have to put your heads back so far? [No.]

3. Why not? [We are looking at something farther away now.]

4. If you put your heads back as before, where would you be looking? [High up in the sky.] This point, i.e. that near objects look higher than distant ones, needs further elaboration by reference to buildings, &c.

5. Does a hill look high if it is very far away? [No.]

Describe the appearance of mountains, which, though far away, are so very high that we have to look upwards to see their summits. Compare with the appearance of clouds, which sometimes have a remarkable resemblance to a range of snow-clad peaks.

6. How long would it take you to go up the highest hill in your district and down again?

Compare with mountains which a man cannot ascend in a day. Tell of Everest and other mountains which no one has ever ascended.

68. Bench-marks

Surveyors have found with accuracy the height above sea-level of certain points. At the spot a Government broad arrow is made in the stone or wood, and on the 6-in. map in the corresponding position is put the sign BM with the height of the arrow in feet.

Take the class to see a bench-mark, or tell them where one is to be found. They are cut deeply in the stone of a wall, gateway, &c. Afterwards let them see the 6-in. ordnance map, and find the height and mark on it.

Let them from the 6-in. map find the heights of a few other places in the district.

A BENCH-MARK

Consider the following questions: —

1. What is the height in feet of the chief hill of the district?

2. What fraction (roughly) is this of a mile?

Tell them that no British hill is a mile high, but that Mt. Everest is about 29,000 ft.

3. How many miles is this? [Nearly 5 ½.]

4. How many times as high is Snowdon (3500 ft.) as your chief hill?

5. How many times as high is Everest (29,000 ft.) as your chief hill?

6. Is the road from A to B downhill or up? (Give places on the map.)

7. Which quarter of a mile of road (1 ½ in. on map) shows the greatest rise in height?

8. How high is your chief hill above the low ground near its base?

Point out that the apparent height of a hill is the height above the place of observation.

69. Determination of Heights—I

Towards noon on a sunny day put in the playground two or three sticks of different lengths in an upright position. Measure their shadows at the same time.

Find the length of one shadow in proportion to the height of the object. Suppose it to be half as long again. Then it will be found that each of the other shadows is half as long again as the object that casts it.

Show from this how to find the height of an object if its shadow can be measured. Put, for example, a book on a wall and measure the distance of its shadow from the foot of the wall. Then from the above the height of the book from the ground should be two-thirds of this distance.

The accuracy of this work will depend on the powers of the class in arithmetic. It may be found best to take a time (in the summer) when the shadows are about equal to the object, and therefore no calculation is needed.

Why must the shadows be measured about the same time? [Because shadows are shortening in the morning and lengthening in the afternoon.]

The objects and shadows should be drawn to scale as shown in figure.

70. The Clinometer

This must be made partly in the woodwork lesson and partly in the drawing lesson. The drawing should be made by all the class; the best attempts may be used and the others rejected. An oblong piece of wood about 10 in. × 15 in. and about 1/2 in. thick must be used. One side AB (fig. 1) must be planed perfectly-straight and level.

A sheet of paper (fig. 2) has a semicircle, radius about 7 in., drawn on it. GO is drawn at right angles to EF. The angles are then marked from O, e.g. 45° half-way from O to E and from O to F, and so on.

The paper is then pasted on ABCD so that EF is parallel to the edge AB.

At G a plumb-line is fixed by boring a small hole through the wood and passing a piece of thread through it. A knot is made on the thread, and at the other end is fixed a weight.

An object is sighted by looking at it along the edge of AB. The plumb-line, of course, remains vertical, and shows the number of degrees from O, which is the angle that AB has been tilted from the horizontal.

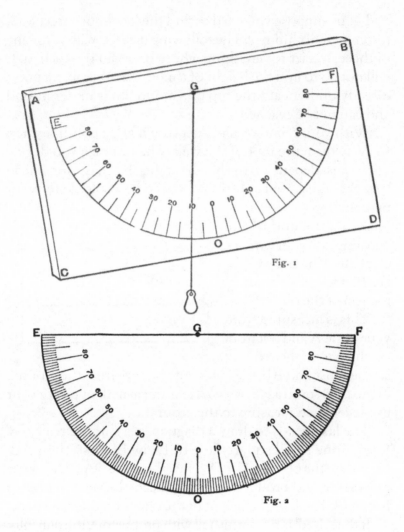

Fig. 1

Fig. 2

71. Determination of Heights— II

Let us suppose you need to find the height of a tree, wall, tower, or mill chimney. The following method will do for any of these, in fact for anything whose base can be reached. It will not do to find the height of a mountain, because a point directly underneath the top of a mountain is underground and can not be reached.

With your chain or tape measure mark a point, say, 100 ft. away from the foot of the wall. The distance you decide upon must depend on the height of the object, e.g. for a small tree or a house wall as little as 10 or 20 ft. would do for your base line.

At A, the end of the base line, use your clinometer to find the angle CAD, i.e. the angle of elevation of the top of the object.

This is measured from your eye at A and not from B, which is on the ground; so you must find AB, the height of your eye from the ground. To do this, stand against a wall and put a mark opposite your eye-level, then measure to the ground.

The last step is to draw a diagram to a suitable scale, say ¹⁄₁₀ in. (one small square) to 2 ft. Draw BE first, then AB. Complete the oblong ABED. Draw the angle DAC with your protractor, and produce AC till it meets the vertical line up from D. CE will be the height of the object on the same scale.

If the school is not furnished with protractors, the difficulty may be got over by moving away from the object till an angle of 45° (or 30°) is obtained. The base line is then measured and the method is as before, except that the set square may be used for the angle.

72. The Water-level

This can be made in the woodwork lesson. A and B are two glass tubes; stout test-tubes with the bottom part filed off will do. C is an india-rubber tube joining them. A straight piece of wood is needed, between 2 ½ ft and 3 ½ ft. long. The other dimensions do not matter. Holes must be made near the ends, through which to fit the tubes. A small hole D may be made in the middle by which to screw it on to a tripod.

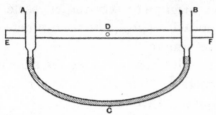

If water is poured into the tubes the level in A will always remain the same as in B, because "water finds its own level". By looking along these two levels, then, we know that we are looking along a horizontal line, even if the wood is not quite horizontal. Red ink or other colouring-matter may be added to the water if the surface is difficult to see.

The instrument should be taken to a fairly level piece of road, and a stick which has a mark on it at the same height as the observer's eye should be viewed. Even a "level" road will generally be found to dip in a distance of 20 yd. or so.

A spirit-level should be made by corking a test-tube of water containing a bubble of air. The bubble will be found to move more freely if methylated spirit is used instead of water. Point out that the bubble always moves to the highest part of the tube. Then show that the tube may be level even if placed on a slanting board. Hence, if we require a surface to be horizontal, we put the spirit-level in two positions, one at right angles to the other, and we lower the side towards which the bubble rises.

73. Gradients

Take the class to a piece of road or other ground that slopes rather sharply, and find the angle of slope as follows:—

A stick must be marked at the level of a boy's eyes. Fix this at the high part of the slope and let the boy stand at the lower part. With the clinometer he must sight the mark on the stick and note the angle. The distance from the boy to the stick must be measured.

Then draw a diagram to a convenient scale, and measure the angle of slope with a protractor or estimate it by comparing with a 30° set square.

If many of the children have bicycles, a table such as that given may be made by them. If not, something similar may be used with reference to carts, &c.

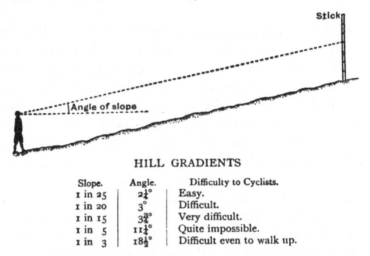

HILL GRADIENTS

Slope.	Angle.	Difficulty to Cyclists.
1 in 25	2¼°	Easy.
1 in 20	3°	Difficult.
1 in 15	3¾°	Very difficult.
1 in 5	11¼°	Quite impossible.
1 in 3	18½°	Difficult even to walk up.

The angle of slope is usually greatly exaggerated. Cyclists usually dismount at an angle of about 3 ½° (1 in 17). (Of course this statement and the others given in the table do not apply to short hills or to very low-geared bicycles.) An angle of 30° on a hill slope has to be scrambled up on hands and knees.

45° is generally referred to as "a vertical precipice" if of a considerable height. An angle of 90° is actually very rare, except for short distances.

Measure the angle of the steepest cliff in the district by means of a clinometer. To be accurate, a point on the top should be observed whose height above the ground is equal to that of the clinometer.

74. Determination of Heights—III

The height of a hill can be found by a method some what like that of Lesson 71.

In this case the measured base line AB will not go right up to the object, but must point straight to it. Two angles must be found—the angle of elevation of H at the nearer end of the base line (B) and at the farther end (A). Of course the angle at B will be greater than that at A, and the base line must be long enough to cause the difference between the angles to be fairly large, say between 15° and 30° difference.

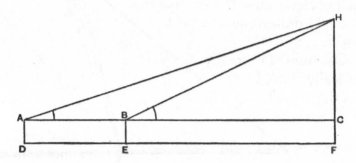

In drawing the diagram, make the rectangle ABED first, just as in Lesson 71. Then draw the two angles, and where they meet at H drop a perpendicular to meet AB and DE produced.

HF is the height of the object to scale.

75. Determination of Heights— IV

Often it is impossible to obtain a suitable base line pointing straight towards the object whose height is to be measured. The following method may then be used, as it affords a valuable exercise in elementary geometry.

Suppose that D is the top of a hill whose height is to be found. C is a point underneath D on a level with the observer, and, of course, is underground.

Measure a fairly level base line AB of suitable length, i.e. not less than about a quarter of the distance to the hill. Put up a pole P vertically, so that DC is in a line with it when viewed from A. Lay a cord 2 or 3 yd. long from A towards P, and another from A towards B. The angle between these cords, i.e. the angle CAB, must then be measured. This may be done by a large protractor made of card board or brown paper. It should be a semicircle of at least 1 ft. radius. The angle CBA must next be found in the same way.

The next step is, by means of the clinometer, to find the angle up to D, i.e. the angle of elevation of D. In an indoor lesson the drawing must be made to scale.

Of course D is really vertically above C, so that the triangle BDC is a ver-

tical one, and cannot be drawn in its true position.

It is necessary to imagine that it is swung round upon BC as a hinge, so as to lie on the paper. CBD¹ is therefore the angle of elevation obtained by the clinometer. DC represents the required height of D above C.

76. Contour Lines—I

Take the class on to a hill-side with the water-level and with a number of poles each about 5 ft. long. Mark these with chalk at the height of one of the boys' eyes. Let this boy use the water-level and others take the sticks. The first boy, by looking along his level, will see whether a certain stick (held vertically, touching the ground) is above or below him. He will direct the holder of it to go up or down the hill till he is at the right level.

In this way the whole string of boys will be standing on the hill-side at the same level. Other boys can be put on the same line by estimation by eye.

Tell the class that the line drawn so as to join the places where the boys are standing is called a contour line. Point out that it is level, and that walking along a contour line means neither descending nor ascending at all.

77. Contour Lines—II

Take the class to a point of known height, e.g. a bench-mark or a place through which passes a contour line marked on the map. It should be a place from whence a view can be obtained, but should not be at the top of a high hill.

Using the clinometer, or the water-level, let the children find places at the same height as the place where they are standing. Let them refer to a contoured map, and find whether they are approximately correct.

It is important to point out that any line joining places of the same height above sea-level is a contour line. Only a few contour lines are shown on a map.

From the same point (or a higher one commanding a better view) let them notice how the streams, where they cross the contour lines, run at right angles to them. Roads and railways, however, run parallel to contours as far as possible. Elicit the reasons.

If a lake, reservoir, or the sea is visible, point out that its shore is a contour line. Imagine its level raised, say, 100 ft. Let the children estimate and, if possible, draw its new shore line. This is another contour line 100 ft. above the other one. Let them say where the new shore line approaches the old one and where the two would recede from one another.

78. Contour Lines—III

A section should now be drawn across a piece of country. Choose as hilly a piece of road as possible near the school. Draw it from the 6-in. map, inserting also the contour lines which cross it. It may be represented as if straightened out, i.e. the copy of the map will not show the bends in the road, but only the true distances from one contour line to another.

Let the class draw a vertical section of the road, using squared paper. The horizontal scale should be the same as the map, i.e. 6 in. to the mile or 1 in. to 880 ft.

The vertical scale will probably have to be exaggerated—the amount of exaggeration will depend on the steep ness of the road. Perhaps 1 in. to 100 ft., i.e. about nine times the horizontal scale, will do. Explain the reason for this exaggeration to the class; on the true scale even a hilly road would look almost perfectly level. Point out that the angles of slope are now not the true ones.

Let the class walk along the road, taking their section with them. They should pace the distances, noting the changes in gradient as they reach them. On arriving at each contour line let them note its direction on each side of the road, according to the map, and compare this with the actual level of the ground.

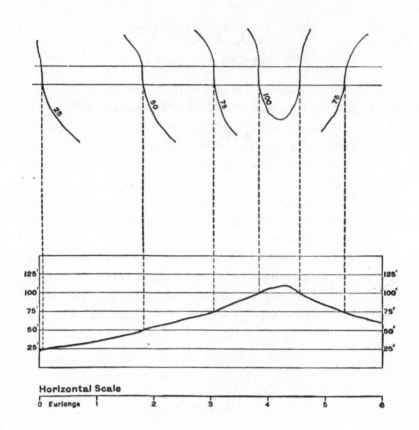

SECTION V
LAND AND SEA

A certain amount of study of nature is an essential part of geography. In some schools the "nature study" taught is almost all Botany, with perhaps a little Zoology. It should be remembered that sky study or Astronomy, weather observation or Meteorology, and Geology are all departments of nature study.

Most of the exercises in this book, in fact, and all those in this part of it, may, if more convenient, be taught in the nature-study lesson.

It is not, of course, claimed that the following suggestions give a complete course of outdoor work in this section. More emphasis will be laid on the part that appeals to the particular district. For example, a school near the sea would greatly extend such work as that given in Lessons 90 and 91.

79. Water

Let the class collect samples of water from various sources, e.g. a reservoir, a spring, a river after heavy rains, a river after dry weather, a stagnant pond, the rain-gauge, &c. Each bottle must be carefully labelled.

Filter the same quantity of each kind of water, using a filter-paper or a circular sheet of blotting-paper. Note which leave a perceptible residue on the paper. Label the filtrates.

Boil away the same quantity of each of the filtrates in an

evaporating-dish. A large amount of each, say about a pint, should be evaporated. Note the amount of residue left in each case. Tabulate the results as follows:—

Source.	Appearance.	Solids Undissolved.	Solids Dissolved.
From rain.	Slightly dirty.	A little—black.	None.
From spring.	Clear.	None.	A great deal.
From swollen river.	Brown.	A great deal.	A little, &c.

Point out that clear water is not necessarily pure. Refer to danger, e.g. from typhoid fever, of drinking water of doubtful purity. The above exercises may be greatly extended by indoor lessons. The difference between the rain that first falls and that which descends after a downfall of some hours may be observed. Very impure water may be distilled. The importance of good water for various industries, e.g. brewing, paper-making, dyeing, laundry-work, may be dealt with.

80. Springs

If possible, take the class to see a spring. If this is impracticable, a few children should visit one and bring an accurate description of it and its surroundings. A sketch map will be useful.

A spring is the place where an underground stream, usually a very tiny one, reaches the surface. Fig. 1 shows the commonest type, on a hill-side, *a* and *b* being pervious beds of rock, e.g. sandstone or limestone, and *c* an impervious one, e.g. clay. Rain percolates through *a* and *b* and flows along the upper surface of *c* to the point where it appears at the surface of the ground.

In another form a pervious bed, *a*, is shut in between two impervious ones, *c*. Water sinks down a, but if it meets a crack in the rocks, such as that shown in Fig. 2, it rises up it and forms a spring. Such a spring, of course, need not be on a hill-side. Usually a whole line of springs will mark where the crack reaches the surface.

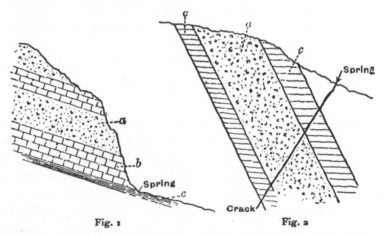

Fig. 1   Fig. 2

If there are any medicinal springs or petrifying springs in the district the water should be evaporated to find the cause of their peculiarities.

STREAM GOING UNDERGROUND NEAR PENYGANT

81. Action of Frost

During frosty weather let the children observe pieces of ice floating on water.

1. Is ice lighter or heavier than water? [Lighter.]

2. About what part of the ice is above water-level ? [One-ninth.]

Most children have very great imaginative powers, and will easily form a correct idea of a huge iceberg from a small block of ice. A tiny paper boat, to represent a ship, will help

the illusion. Refer to dangers to navigation. Show how soil
and pebbles on the ice may be carried to a distance, and, when
the iceberg melts, will fall to the bottom of the sea. Refer to
the Banks of Newfoundland, partly formed in this way.

Take a stout glass bottle, fill it with water, cork it firmly and
tie the cork on, taking care that the bottle is quite full. Place
it outside during a sharp frost. Observe it when it has burst.

3. Why do water pipes burst in winter ?

4. Why is the burst generally not noticed till the thaw?
[The ice stops the leak.]

During a thaw, visit a ploughed field or a garden which
has been roughly dug.

5. What change is there in the soil? [It readily crumbles.]

6. Why is this? [Ice formed in the cracks.]

Speak of the value of this action to the farmer. If possible,
let observations be made of a rock surface during a thaw.

7. When are blocks of stone most likely to fall from a cliff?
[During thaw.]

High cliffs often have notices, "Beware of falling rock",
at such times.

82. Rivers—I

Let the class measure, by pacing, a distance of, say, 10 chains
(1 furlong) along a river bank. Throw some conspicuous object
that will float into the water, and time its passage along the
measured distance. Calculate the speed in miles per hour.

Go to the nearest bridge and, by means of a string and a
heavy weight, measure the depth at, say, every yard along the
bridge. Draw a vertical section of the river on squared paper.
The scale might be $\frac{1}{10}$ in. represents 6 in.

Count the number of squares in the irregular figurAGHKLF,
neglecting part squares that are less than half, and counting

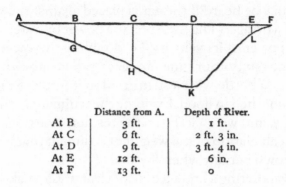

	Distance from A.	Depth of River.
At B	3 ft.	1 ft.
At C	6 ft.	2 ft. 3 in.
At D	9 ft.	3 ft. 4 in.
At E	12 ft.	6 in.
At F	13 ft.	0

as whole ones those greater than half. Suppose there are 91. Four squares represent a square foot. Therefore the area of cross-section is 22 ¾ sq. ft.

Suppose the speed of the stream is 2 ¼ miles per hour, or 11,880 ft. per hour. Then the number of cubic feet of water passing under the bridge per hour will be

$$9\frac{1}{4} \times 11 \times 11,880 = 270,270$$

As 36 cu. ft. weigh about a ton, there will be, roughly, 7500 tons of water per hour passing this place.

83. Rivers—II

Look at the stones at the bottom of a shallow stream when the water is clear.

1. What shape are the stones? [Rounded and smooth.]
2. Are they moving? [No.]

Mark a few of the stones and replace them. Some of the children should be told to observe them every few days and report when they have moved down-stream. The slow movement of loose material along the bed, moving only in flood-time, will thus be shown. Sometimes the rattling of the stones may be heard if the ear is placed against the wood or stone of a bridge. (The transporting power of water varies as the 6th power of its velocity, i.e. doubling the speed means that stones sixty-four times as heavy can be moved.)

Throw a bottle with a string tied to it into the middle of the stream when in flood. Examine the sediment in the water. Filter, dry, and weigh it. From the results of Lesson 82 make a rough calculation of the weight of sediment that your river brings down per day when flooded.

By evaporating a known bulk of clear water, make a rough calculation of the weight of dissolved matter brought down. Frequently a tiny side stream will build out a delta into a quiet part of the main stream. Such a delta may be found in a pool in an uneven road. It should be examined to show the bedding, the coarsest material being underneath and the finest at the top.

84. Waterfalls

In deciding which waterfall to visit, remember that the points can be observed quite as well in a small fall as in a large one. Pictures of a larger fall, such as Niagara, will help considerably. Approach the fall from below. Point out the gorge. Let questions be considered such as the following:—

1. What height are the banks above and below the fall?
2. Are all the rocks on the face of the fall alike? [No.]
3. Where is the softer one? [At the base.]

Show how this is being worn away more quickly than the upper layers. Tell the children that in a very large fall people can walk between the cliff and the water.

4. How is this? [The upper rock overhangs.]
5. What will happen to this rock in time? [It will fall for lack of support.]

6. What evidence is there that this has happened before? [Masses of hard rock in the stream below the fall.]

7. What effect will this have on the position of the water-fall? [It moves up-stream.]

Mention that Niagara has been shown by measurement to move about 5 ft. per year in this way. Point out how a gorge is thus formed.

8. Will the gorge become longer or shorter? [Longer.]

The gorge below Niagara is about 7 miles long. The time the river has been flowing may be roughly estimated by dividing 7 miles by 5 feet.

9. How has the deep pool just below the fall been made? [By falling water aided by stones and gravel.]

85. A Rainbow

Such a beautiful and striking phenomenon as a rainbow should not be allowed to pass unremarked because the children are not old enough to fully understand a lesson on the spectrum.

On a sunny day they may place a three-cornered piece of glass in the full light and notice the colours that are cast. If a fountain can be visited they should go round it till they see a rainbow-like appearance.

How must we stand to see this? [With our backs to the sun.] Tell the children that a rainbow is always formed when the sun shines on falling drops of water. If it rains and the sun shines at the same time the bow is sure to be seen, but always on the side farthest from the sun.

There must be both rain and sunshine where the bow is, but not necessarily where the observer is standing.

A simple talk about white light being composed of "all the colours of the rainbow" may follow. A circular card may be coloured in sectors, respectively red, yellow, and blue, and spun round by means of a top. The card will look white, or rather grey owing to impurity of all colouring matter. Encourage the children to observe carefully the next rainbow they see and be prepared to answer the following questions: —

1. What colours were there, and in what order? [Red, orange, yellow, green, blue, violet.]

2. Was a second bow visible beside the bright one? [Usually visible.]

3. Was it higher or lower than the bright one? [Higher.]

4. Which was the top colour in the bright one? [Red.]

5. Which was the top colour in the faint bow? [Violet.]

My heart leaps up when I behold
A rainbow in the sky.

Wordsworth.

Triumphal arch, that fill'st the sky
When storms prepare to part,
I ask not proud Philosophy
To teach me what thou art.
How glorious is thy girdle cast
O'er mountain, tower, and town,
Or mirrored in the ocean vast
A thousand fathoms down!

Campbell

86. Rocks—I

Visit some section through the earth, such as a quarry, railway cutting, or cliff. Elicit answers to the following questions:—

1. About how deep is the soil from the surface? [It varies in different localities from nothing at all to a depth of several feet.]

2. Is the rock all in one piece? [No.]

3. How do the chief cracks run? [Nearly horizontally (generally).]

Explain the term "bed" or "stratum". Often one bed may easily be traced owing to its distinct colour, hardness, or softness.

Encourage the children to bring specimens of local rocks. Two distinct schools may exchange specimens, and the following, at least, should be obtained: Gravel, sand, clay, sandstone, shale, slate, limestone, basalt, chalk, granite, marble. The last two may be obtained from a monumental mason. Examine the specimens. Note their hardness by trying to scratch each with a knife.

Let the children make observations during their rambles

which will enable them to answer some of the following questions: —

1. Of what rock are the hills near here composed?
2. What rock is found in the valleys?
3. What rocks have you seen forming a cliff?
4. Was it vertical or sloping? Why? [Only very hard rocks will form nearly vertical cliffs.]
5. Are stones or bricks used for local buildings? Why?
6. Are slates or tiles used for roofing? Why?
7. Are stones, fences, or hedges used for dividing fields?
8. What ornamental stones have you seen in the walls or pillars of banks, churches, &c. ?

87. Rocks—II

If possible, a visit should be made to a limestone or chalk district. The typical features of the scenery of such a district should be pointed out. A lime-kiln should be seen by the children.

Collect specimens of limestone and chalk and compare them. Chalk is really a soft variety of limestone. Note the effect of water on them: it is absorbed but no further change occurs.

Put a piece of the rock (about the size of one's fist)into a hot fire for about two hours. Allow it to become cold and then test it by pouring water on it. Take care to distinguish carefully between limestone (a natural substance), quicklime, and slaked lime (both artificial). As lime never occurs in nature it is obviously incorrect to speak, for example, of a spring or river having *lime* dissolved in it.

Let the children mix some slaked lime with ground-up cinders to illustrate the making of mortar. It should be placed outside to set.

Of course valuable science lessons may also be given upon

limestone and lime to illustrate the meaning of chemical action, the formation of heat, and to lead to consideration of
carbon dioxide.

Take a specimen of sandstone and powder it in a mortar to
show that it is consolidated sand. If the grains are coloured,
heat them with hydrochloric acid to show that the colouring
is only on the surface, and will dissolve leaving white sand.

Let paving-stones be examined. The layers of mica will
be seen which will enable the stone to split easily along certain lines. Broken across, the stone will be seen to be mainly
composed of sand grains.

An elementary book on Geology will give suggestions for
further work of this nature.

FOLDED BEDS OF MOUNTAIN LIMESTONE, DRAUGHTON, NEAR SKIPTON

88. Minerals

By a mineral a geologist means a substance of definite
composition forming part of a rock, e.g. quartz, iron-ore. The
term, however, is generally extended to include also any rock

which has a market value, e.g. limestone or coal.

Take a rock, such as granite, and show the minerals in it, quartz, looking like glass, mica, in thin brown or black flakes, and felspar, dull and variously coloured.

Let a collection of minerals from the district be made. Make it clear that such substances as iron and lead are never found in the earth. What is found is a mineral (ore), from which the metal may be obtained by suitable chemical means. If, for example, iron-ore is found near, a lesson on the blast furnace may follow.

Let the following questions be answered after suitable observation.

1. What mines, quarries, or pits are there in the school district?

2. What are the prices of minerals on the spot, e.g. of coal at a pithead?

3. What are the prices delivered to the consumer?

4. What minerals are needed for local industries, e.g. clay for brick-making?

5. What minerals do you see transported by rail, canal, &c.?

6. Between what places are they being carried?

7. Ask for list of metal objects, e.g. lead pipes, iron gas-lamps, copper telephone wires, steel rails, &c.

This will lead to a consideration of where such things are made, and where the mineral comes from.

89. Soils

Visit a section as for Lesson 86, but with special reference to the soil. Let the class make a drawing so that they will note:

1. That the lower beds are little broken.

2. That higher up the horizontal vertical cracks are more marked, so that oblong pieces are formed.

3. That higher still there are smaller oblongs, with corners rounded off.

4. That near the top these merge into soil.

Point out the action of frost when breaking up the rock.

Often roots of trees may be seen penetrating the cracks and helping the action. The solvent action of water may be referred to. Break a surface rock and show how different the inside is from the surface, which has been exposed to the weather. Refer to soil being made finer by passing through the bodies of earthworms.

Let specimens be obtained of several kinds of soil. Each should be closely examined. Bits of rock will be found which will show from what formation the soil was derived. Point out the bits of leaf and twig and other signs of plant life. Burn these off with a Bunsen burner, and show that what is left is only finely broken rock. A rich soil will lose more by this treatment than a poor one. Tabulate the soils somewhat as follows: —

Locality.	Colour.	Loss when 100 parts by weight are burned.	Fineness.
Quaker wood	Nearly black	23 parts	Very fine
Near sand-pits	Dark yellow	5 parts	Coarse—many small pebbles

90. Sea-water and Waves

Evaporate some sea-water and compare the residue with those left by the specimens of water in Lesson 79. If possible, take specimens from a tidal river at low water, and at high water, and at an intermediate time.

The children should be told that all river water contains a little salt and a little limestone. (See Lessons 83 and 87). This water reaches the sea and, when the water evaporates there, the solids are left behind. We should therefore expect sea-water to contain much salt and much limestone. The lime-stone, however, is constantly being removed by the growth of creatures such as the oyster, the coral, &c., whose shells are made of limestone. (Carefully guard against the peculiar error of calling the coral an "insect", and the equally com-mon mistake of saying that the coral "builds" a reef. Actually the coral is an animal, but it may be called "shell fish" like an oyster or a crab. The only "building" it does is to die and leave its shell to form part of a deposit along with millions of others which may together form an island.)

As a result, we find that sea-water contains much salt and a little limestone. There are also other substances which cause its bitter taste.

Point out that if part of the sea is cut off from the land, and

COAST DENUDATION NEAR DAND-LE-MERE, HOLDERNESS

climate causes the water to dry up, a bed of salt will result. This may be covered by sand blown by the wind, and the result will be beds of sandstone and salt, as in Cheshire, &c.

Observations on waves can be made on any sheet of water. The height of waves should be estimated by comparison with objects such as stones on the beach. The great exaggeration of such expressions as "mountains high" should be pointed out. It is doubtful whether in the worst storm the height ever exceeds 50 feet, and waves even of 20 feet are rare.

By throwing a piece of wood in the water show that the movement of a wave is only up and down, except near shore, where the wave breaks.

91. Tides

It is not advisable in the elementary school to attempt to explain the causes of tides. It will be quite sufficient to give a

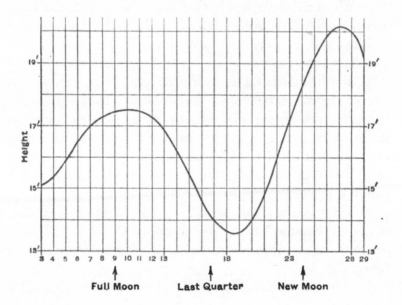

clear idea of what the movements are, and to show that they are connected with the moon.

In an inland school observations can be made during the holidays by a certain number of the children. Questions such as the following will show what should be looked for:—

1. Observe the time when the water has gone down to its lowest point, and "low water".

2. How long after this was the water highest? [A little over six hours.]

3. What distance has the water advanced up the beach in this period?

Point out that where the beach slopes very steeply there is little advance, perhaps only a few feet, whereas on a nearly level shore there may be a forward and backward movement of a mile or more.

4. What time elapses between high tide and the next high tide? [A little over twelve hours.]

The time of rising or of setting of the moon should be found for a few days, either by observation or from tables.

5. What time elapses between moonrise and the next moonrise? [A little over twenty-four hours?]

The children may now be told that the tides are caused (chiefly) by the moon.

Either from tide-tables or from direct observation let the level of high water be found for a period of at least a fortnight. Let a graph be drawn similar to the one shown, and mark on it the times of full, new, and quarter moon. Answers may then be obtained to the following questions:—

6. When do the highest high tides occur? [Just after full and new moon.]

7. When do the lowest low tides occur? [At the same times.]

Give the name "spring" tides applied to these, and point out that the word has no connection with the season named "spring".

8. When is the least change of level in the tides? [Just after

quarter moon.]

These are the "neap" tides.

9. Are all spring tides of equal height? [No.]

The reasons for this are very complicated. Children in coast towns may learn something of them from sea men.

There twice a day the Severn fills;
The salt sea-water passes by,
And hushes half the babbling Wye,
And makes a silence in the hills.

Tennyson.

Evening Tides at Liverpool

Date.	Height.	Date.	Height.
December 3	15 ft. 1 in.	December 17	14 ft. 0 in.
,, 4	15 ft. 4 in.	,, 18	13 ft. 7 in.
,, 5	15 ft. 11 in.	,, 19	13 ft. 7 in.
,, 6	16 ft. 5 in.	,, 20	14 ft. 0 in.
,, 7	17 ft. 0 in.	,, 21	14 ft. 11 in.
,, 8	17 ft. 3 in.	,, 22	16 ft. 1 in.
,, 9	17 ft. 6 in.	,, 23	17 ft. 3 in.
,, 10	—	,, 24	18 ft. 3 in.
,, 11	17 ft. 6 in.	,, 25	19 ft. 0 in.
,, 12	17 ft. 6 in.	,, 26	19 ft. 11 in.
,, 13	16 ft. 9 in.	,, 27	20 ft. 2 in.
,, 14	16 ft. 2 in.	,, 28	20 ft. 0 in.
,, 15	15 ft. 5 in.	,, 29	19 ft. 2 in.
,, 16	14 ft. 7 in.		

SECTION VI
HUMAN GEOGRAPHY

The most important part of geography is that which deals with the activities of man. Physical geography is in a sense only preliminary to this, for climate, relief, &c. have little interest except in so far as they affect the peoples of the earth.

We want to broaden the child's outlook by leading him to understand the lives of others. By degrees we should enlarge his knowledge and sympathy from the narrow circles of family and school until he realizes that he is a member of a village or town community, a citizen of Britain, and a subject of the Empire. Finally, he should recognize that he is dependent upon, and has duties towards, the peoples of all the countries of the earth. In this way he is being prepared for his responsibilities as a citizen.

Such an enlargement of social knowledge must be based on the small circles of people with whom he can come into direct contact. The following lessons are given merely as suggestions —they should be greatly enlarged and amplified.

92. Roads

Examine (a) a well-made road, (b) a neglected country lane, preferably after heavy rain. Consider the following questions:—

1. How does a well-made road slope in a direction at right angles to its length? Use a spirit-level (Lesson 72).

2. How many puddles are there in a certain length of each? How many loose stones are there in a certain length of each?

3. What is the depth of the deepest rut you can find in each?

4. Let certain children watch an important road for twenty minutes, and keep a record of the number of (a) pedestrians, (b) cyclists, (c) light carts, (d) wagons, (e) motor-cars, (f) motor-lorries that pass. Let them find, if possible, what goods are being carried along it, where they come from, and where they are going. The increasing use of roads for transport of heavy goods should be mentioned. By the aid of the 6-in. or 1-in. map, discover to what extent the roads follow the contour lines. Discover the greatest gradient on a main road, and compare it with the greatest railway gradient. Note some decided bends on such a road, and account for them.

Roads tend to radiate from towns. Elicit this fact from direct observation, and illustrate it from maps.

A Saturday ramble becomes much more interesting if it is first traced out on a 6-in. map. Short cuts are often shown which would otherwise be missed.

It must be remembered that a footpath marked on an ordnance map is no evidence of right-of-way. Children should be warned of the importance of closing gates, of keeping out of hay-fields in May and June &c.

Give to me the life I love,
Let the lave go by me,
Give the jolly heaven above
And the by-way nigh me,
Bed in the bush with stars to see,
Bread I dip in the river.

Give the face of the earth around
And the road before me.

Stevenson.

93. Railways

Let the children make a map of the railways for a radius of 10 miles or so from the school. It should be corrected from the 1-in. ordnance map. Let a table be made as follows, A being the local station: —

Distance from A by Road.	Distance from A by Rail.	3rd Class Fare from A.
B C &c.		

Set questions such as the following, giving a week or two for the answers to be discovered:—

1. What railway company owns your railway?

2. Describe how the engines and carriages of this company are coloured.

3. What other carriages and engines pass through A?

4. About how many (1) passenger, (2) goods trains pass A per day?

5. What goods have you observed in a local goods train?

6. What is the fare to X? (A large well-known town near.)

7. What is the best train to X in the day?

8. What is the average speed of this train?

9. Name some distant towns you can reach from A without changing trains?

10. Name some villages without a station. Why have they no station, &c.?

From the 6-in. map show how the railway follows the contour lines. Explain the meaning of the gradient posts along the line. If there is a steep gradient near, the children will observe the difficulty of a heavy train in getting up it.

As we rush, as we rush in the Train,
The trees and the houses go wheeling back,
But the starry heavens above the plain
Come flying on our track.

J. Thomson.

94. Canals

Let a map be drawn showing the course of any canals near the school. Give opportunities for observation to consider the following questions: —

1. What depth has the canal at a certain bridge? [Measure by weight at the end of a string.]

2. Which is it important to know, the least or greatest depth? [The least depth determines the size of barges using it.]

3. Can two barges pass each other everywhere?

4. How does the water differ from that in a river?

5. Is every part of the canal at the same level?

Let a diagram be made of a lock, with an explanation of what happens as a barge passes through. Mention the need for a water-supply, and the reservoirs that give it.

6. How are the barges moved? [Refer to horses, sails, tugs, and engines.]

7. Can a man move one? [Yes, very slowly.]

Speak of the reasons why big loads can be easily moved—absence of friction, no gradient, support of the water. Explain cheapness of carrying heavy substances.

8. Make a list of substances you have seen on canal barges.

9. Is fruit conveyed this way? Why not?

10. Time a barge over a measured distance, and compare speed with that of a train.

Speak of the effect that rapid movement would have in washing down the banks.

11. Why is china and glass often carried by canal? [Freedom from shaking.]

12. Give all the advantages and disadvantages of canal transport.

95. Local Town Industries

The fact should be elicited by questions that all commodities, if traced back far enough, come from the earth. Scarcely any, however (coal is an exception), are capable of being used till they are changed. Hence we get the two great classes of industry: (1) extractive, e.g. mining, agriculture; (2) manufacturing, e.g. weaving, iron smelting. Other classes are (3) trade and commerce; (4) building and construction; (5) transport; (6) services to others, e.g. the doctor and the teacher.

Make a list of typical local workers and find out in each case the following information:—

1. How many hours does he work per day?

2. Is the work interesting, very hard, dangerous, &c.?

3. Did he have to work long with little pay to learn the work?

4. About what wage does he get?

5. Is there much danger of unemployment for long periods?

6. What prospect of promotion is there?

At least a year before a child leaves school he should be encouraged to decide what he would like to do. This would greatly reduce the number of youths who go into "blind alley" occupations.

Questions should be given on the chief works of the districts.

1. Who are the chief employers of labour?

2. How many people does each employ?

3. What does each mill or factory produce?

4. Where are the products sent?

5. What raw material is needed and whence does it come? &c., &c.

96. Local Farming

If the local farms are nearly all in pasture, it will be advisable to compare notes with an agricultural district, and conversely.

Allow observations to be made for the following: —

1. Name one of the chief farmers of the district.
2. Give the number of his fields.
3. Estimate the area of his land.
4. About how many animals does he keep?
5. Give a list of materials he needs, e.g. cattle food, &c.
6. Say as exactly as possible whence he obtains them.
7. What products does he sell?
8. Where does he sell them?
9. Give some approximate prices.

Using a 6-in. map of a farming area, colour the fields according to their use, e.g. (a) for pasture, (b) for hay, (c) for roots, (d) for corn, &c.

Observations should be made to show the effect of altitude on crops. For instance, compare two farms, one much higher than the other, in the following way:—

Crop.	LOWDALE FARM. Date (first cut).	HIGHDALE FARM. Date (first cut).
Grass	30th June.	5th August.
Wheat	15th August.	None grown.
Oats		
Barley		

In a city school the best plan may be to start from the products and trace them to their origin. For example, the children may be asked who delivers their milk. If it comes by train, the station it is dispatched from may be found, and some at least of the children may visit the village and find the dairy farms. The market will also be a starting-point for similar work.

97. Pleasure and Health

It is important to remember that man has other activities besides those connected with money-getting.

Ask questions such as the following:—

1. Give the outdoor pleasures of your district. (Playing football, watching football, taking walks, &c.)

2. Arrange these in order, based on the number of people doing each on a fine day.

3. Arrange them in order of cost.

4. Give the effect of change of seasons upon these pleasures. This will lead to talk about other countries, e.g. (1) where skating and tobogganing are very common; (2) where hunting and shooting are common; (3) where the heat prevents most outdoor games.

5. Name the largest buildings devoted to indoor pleasure.

6. Name local districts whose inhabitants cannot well visit these.

The advantages of country life for outdoor pleasure and of town life for indoor pleasure may be brought out. The influence of each kind of pleasure on health can be estimated.

7. Give a list of places visited for pleasure or health. Point out the large number of people who obtain their living by providing for the wants of visitors to towns like Blackpool and Brighton, and to countries like Switzerland.

8. Are there any places near the school which are visited for health?

9. What places have you known sick people from your district to visit?

The special value of sulphur springs (Harrogate, &c.) for skin diseases, also of salt baths for rheumatism (Droitwich), of a mild winter for delicate persons (Torquay, &c.), may be pointed out.

98. Historical Geography

The work done in this subject will vary very much according to the position of the school. In some localities many lessons may profitably be given, in others only a few. In most places, however, there is a ruined castle or abbey, a Roman wall or camp, or at least an old church. The work when visiting the place belongs to the subject of history, but often connected work in geography is useful. Such questions as the following may be considered.

1. Why was the castle or road built here?

2. Does it show any change in the distribution of population?

3. Draw the contours round a castle or camp. The building is often greatly affected by them.

4. Describe the changes in the local geography since the building was used.

In some cases even the physical geography, and in all cases the human geography, has considerably altered.

Very often the 6-in. map will yield many place-names which are suggestive. Information may be collected from aged people, and from books, about the state of the district many years ago.

99. Local and State Government

Each of the many activities of the municipal or county council should be dealt with. The property of such public bodies is generally marked with certain letters, and the meaning of these contractions should be known. When an excavation is made in a road, attention may be called to the drains, electric cables, gas and water pipes, &c., disclosed. The overhead system of telegraph and telephone wires and electric-light cables should be shown, and the difference between the kinds of wires used may be indicated. Questions such as the following are suggested:—

1. From whence does the school water come? The collecting and storage reservoirs should be found on the 1-in. map.

2. How do your parents pay for the water used in your house?

3. Where is the nearest fire-station to your school? your house?

4. How could you most quickly send warning of a fire at each?

5. Where are the gas-works supplying your school?

6. Are they near a railway siding? Why?

7. Where are the sewage-works nearest your school?

8. What work is done there?

Similar questions may be put in reference to electric light, trams, parks, schools, libraries, the police force, roads, &c.

9. Who manages the above concerns and appoints and pays the people employed?

10. Give the names of some of the members of this body?

11. How is this body elected?

12. How do they get the money that they have to spend?

The importance of all the kinds of work done and the large sums expended should be pointed out. Hence follows the importance of elections of such public bodies.

State Government may be treated in a similar way. Reference may be made to the letters G.R. (e.g. on a post-office van) and the Government broad arrows (e.g. on a bench-mark or on convicts' clothing) as signs of State property. The post-office, courts of law, prisons, and fighting forces may be dealt with. This will lead to consideration of Parliament, elections, and laws.

100. Comparison with other Districts

A great amount of real geography can be learnt by corresponding with some other school, preferably one at a distance, in a different climatic region, and with different industries round it. The following questions might be asked in the letter, others will no doubt suggest themselves: —

Physical
1. What is the highest hill within walking distance of your school?
2. How long does it take to climb?
3. How wide is the river nearest your school?
4. Are there fish in it? If so, what fish?
5. Can boats be used on it? If so, what size of boat?&c.

Climatic
6. By what time is it too dark to read outside at midsummer? at midwinter?
7. Did you have skating last winter? How many days?
8. Will wheat grow near you? If not, do you know why not? &c.

Industries
9. Do most of the people near you work outdoors or indoors?
10. Is there one very important kind of work? If so, what?

11. On the loneliest road near, what distance would you go without passing a house? &c.

Miscellaneous

12. Do any of the people near you talk in a way we, living at a distance, should not understand? Can you write down an example?

13. Do you eat any food which is not common all over Britain, &c.

Photographs should be obtained and exchanged for others, so that by degrees a collection may be obtained illustrating various geographical features. The examples here given show the unequal denudation of one of the Bridestones, near Pickering, and a coast scene near Flamboro' Head showing the steep slope of the chalk and the gradual slope of the softer boulder clay above.

BRIDESTONE NEAR PICKERING

CLIFFS NEAR FLAMBOROUGH HEAD

APPENDIX I
OTHER USEFUL APPARATUS

In some cases it may be advisable to buy apparatus from makers of scientific instruments. The following will be found extremely useful:—

1. Rain-gauge (British Association pattern)
2. Maximum thermometer
3. Minimum thermometer
4. Plane table and tripod
5. Spirit-level
6. Trough compass
7. Clinometer and sight rule
8. Surveyor's chain (66 ft.)
9. Linen tape rule (66 ft.)
10. Surveying rods (half-dozen)
11. Theodolite
12. Cross-head and rod

An astronomical telescope is an expensive piece of apparatus. By means of two lenses of differing focal length, however, many valuable observations may be made. The first, held close to the eye, should be of focal length 4 in.; the second, held about a foot away, should be of focal length 8 in. Of course the image seen is inverted, but that does not matter when viewing the moon or stars.

APPENDIX II
MAPS

The Ordnance survey maps issued by the Government are by far the best. They may be obtained through a stationer or direct from the Ordnance Survey Offices, Southampton.

1. The most useful for little children is on the scale of $1/2500$ (or about 25 in. to the mile. It is more of the nature of a plan than a map, showing every house and garden, the areas of fields, &c.

2. A larger area is shown in the 6-in. maps (3 ml. by 2 ml.). These will show, for example, the whole of a good-sized town, every street being marked. Bench marks and contour lines are given.

3. The 1-in. large-sheet series cover a large district (27 ml. by 18). They are printed in colours—hills brown, woods green, water blue, contours red. These are particularly useful for children who can walk a good distance or cycle. The characteristic sheet should certainly be obtained.

4. The 1/2-in. map on the layer system is a beautiful map, altitudes being shown by colouring of different shades. It is invaluable in teaching contours.

5. The 1/4-in. map in colour covers an area of about one-tenth of England.

APPENDIX III

TABLE OF DECLINATION OF SUN

The declination of the sun for a given date is the latitude where the sun is vertical at noon on that date.

The following declinations are not exact, but are near enough for school purposes. The values for intermediate dates can be obtained by simple proportion.

Jan. 1.	23° S.	Feb. 1.	17¼° S.	Mar. 1.	8° S.
„ 8.	22¼° S.	„ 8.	15¼° S.	„ 8.	5¼° S.
„ 15.	21¼° S.	„ 15.	13° S.	„ 15.	2½° S.
„ 22.	19¾° S.	„ 22.	10½° S.	„ 22.	0
April 1.	4¼° N.	May 1.	14¾° N.	June 1.	22° N.
„ 8.	7° N.	„ 8.	16¾° N.	„ 8.	22¾° N.
„ 15.	9½° N.	„ 15.	18¾° N.	„ 15.	23¼° N.
„ 22.	12° N.	„ 22.	20½° N.	„ 22.	23½° N.
July 1.	23¼° N.	Aug. 1.	18¼° N.	Sept. 1.	8½° N.
„ 8.	22½° N.	„ 8.	16½° N.	„ 8.	6° N.
„ 15.	21¾° N.	„ 15.	14¼° N.	„ 15.	3¼° N.
„ 22.	20½° N.	„ 22.	12° N.	„ 22.	0
Oct. 1.	3° S.	Nov. 1.	14½° S.	Dec. 1.	21¾° S.
„ 8.	5½° S.	„ 8.	16¼° S.	„ 8.	22½° S.
„ 15.	8¼° S.	„ 15.	18¼° S.	„ 15.	23¼° S.
„ 22.	10¾° S.	„ 22.	20° S.	„ 22.	23½° S.

APPENDIX IV
SOLAR AND CLOCK TIME

This table shows the number of minutes that the sundial is fast or slow by Greenwich time. It will also show the number of minutes that noon by the sun is before or after the clock noon.

Of course the longitude of the place will cause another difference between local and clock time, which can be calculated by remembering that every degree we go west of Greenwich makes local noon come four minutes later in the day.

Jan.	1.	$23°$ S.	Feb.	1.	$17\frac{1}{4}°$ S.	Mar.	1.	$8°$ S.
„	8.	$22\frac{1}{4}°$ S.	„	8.	$15\frac{1}{4}°$ S.	„	8.	$5\frac{1}{4}°$ S.
„	15.	$21\frac{1}{4}°$ S.	„	15.	$13°$ S.	„	15.	$2\frac{1}{2}°$ S.
„	22.	$19\frac{3}{4}°$ S.	„	22.	$10\frac{1}{2}°$ S.	„	22.	0

April	1.	$4\frac{1}{4}°$ N.	May	1.	$14\frac{3}{4}°$ N.	June	1.	$22°$ N.
„	8.	$7°$ N.	„	8.	$16\frac{1}{2}°$ N.	„	8.	$22\frac{3}{4}°$ N.
„	15.	$9\frac{1}{2}°$ N.	„	15.	$18\frac{3}{4}°$ N.	„	15.	$23\frac{1}{4}°$ N.
„	22.	$12°$ N.	„	22.	$20\frac{1}{2}°$ N.	„	22.	$23\frac{1}{2}°$ N.

Example.—To find local noon at a place 1 ½° W. on 1st October. 1 ½° makes a difference of 6 minutes later than local noon at Greenwich. But by the above table local noon at Greenwich is 10 minutes before 12 o'clock.

Therefore local noon at our position is 4 minutes before 12, i.e. at 11.56 a.m. Greenwich time.

APPENDIX V
MAGNETIC DECLINATION

By this is meant the angle between the geographical north-south line and the direction of the compass needle. In other words, it is the angle between the geographical meridian and the magnetic meridian.

Note that the compass needle does not point (as some books state that it does) to the magnetic pole. It is therefore incorrect to state that declination is the angle between lines drawn to the geographical pole and the magnetic pole respectively.

The declination at London is at present (1919) about 14 1/4° W. On a line from Newcastle to Plymouth it is about 1 1/2° more, at Dublin 3 1/2° more, and at Dover 1/4° less. Everywhere in the country it is decreasing at the rate of about 1/6° per year.

Made in United States
North Haven, CT
19 July 2023